ANGULAR MOMENTUM IN GEOPHYSICAL TURBULENCE

Angular Momentum in Geophysical Turbulence

Continuum Spatial Averaging Method

By

Victor N. Nikolaevskiy
Schmidt's United Institute of Earth Physics,
Russian Academy of Sciences, Moscow

KLUWER ACADEMIC PUBLISHERS
DORDRECHT / BOSTON / LONDON

A C.I.P. Catalogue record for this book is available from the Library of Congress.

ISBN 978-90-481-6478-3

Published by Kluwer Academic Publishers,
P.O. Box 17, 3300 AA Dordrecht, The Netherlands.

Sold and distributed in North, Central and South America
by Kluwer Academic Publishers,
101 Philip Drive, Norwell, MA 02061, U.S.A.

In all other countries, sold and distributed
by Kluwer Academic Publishers,
P.O. Box 322, 3300 AH Dordrecht, The Netherlands.

Printed on acid-free paper

CONTENTS

PREFACE

Turbulence theory is one of the most intriguing parts of fluid mechanics and many outstanding scientists have tried to apply their knowledge to the development of the theory and to offer useful recommendations for solution of some practical problems. In this monograph the author attempts to integrate many specific approaches into the unified theory.

The basic premise is the simple idea that a small eddy, that is an element of turbulent meso-structure, possesses its own dynamics as an object rotating with its own spin velocity and obeying the Newton dynamics of a finite body. A number of such eddies fills a coordinate cell, and the angular momentum balance has to be formulated for this spatial cell. If the cell coincides with a finite-difference element at a numerical calculation and if the external length scale is large, this elementary volume can be considered as a differential one and a continuum parameterization has to be used.

Nontrivial angular balance is a consequence of the asymmetrical Reynolds stress action at the oriented sides of an elementary volume. At first glance, the averaged dyad of velocity components is symmetrical,

$$<u_i u_j> \equiv <u_j u_i>$$

However, if averaging is performed over the plane with normal n_j, the principle of commutation is lost. As a result, the stress tensor asymmetry

$$<\rho u_i u_j>_j$$

is determined by other factors that participate in the angular momentum balance. This is the only possibility to determine a stress in engineering.

Of course, the additional balance has to include a new kinematics' variable. This is the spin velocity of a sub-scale turbulent eddy.

A fully adequate description needs the methods of generalized continuum mechanics. To study the corresponding philosophy, for 15 years I took part in the famous seminar, led by Leonid Sedov in Moscow State University, and spent half of 1974 as a post-doctor with Clifford Truesdell at Johns Hopkins University (Baltimore). These professors were leaders in the field of applied mathematics and paid essential attention to the renovation of the Cosserat ideas as well as their applications to wide scope of problems. The inspiration that I received from them was great.

I work in the Institute of Earth Physics of the Russian Academy of Sciences mainly on the problems of rock/soil dynamics*, and I have had many opportunities, besides contacts with Moscow professors, to visit the US and meet many distinguished researchers and academics. Discussions with them intensified my studies and in general supported my approach but some aspects were argued that gave additional impetus for further research. Some extremely difficult questions received proper answers only many years later. Some objections raised were connected with misleading statements contained in even well known publications. Only after satisfactory answers were received, did I decide to write this book.

L. I. Sedov and P. Y. Polubarinova-Kochina had submitted my papers to the Proceedings of the USSR Academy of Sciences (DAN). George Batchelor, interested in dynamics of suspensions at that time, invited me to visit Cambridge University; A. N. Kolmogorov understood immediately my proposal and gave his approval. Stanley Corrsin and Laszlo Kovasznay spent many hours with me discussing the problem and recommended that I consider phenomena in which conventional approaches had failed.

The tornado problem was selected because there is a sharp front between a turbulent object and an ambient laminar atmosphere that cannot be described by smooth solutions of the Navier-Stokes equations even by viscosity modifications suggested by L. Prandtl et al. It is known that such object as the "fire ball" of atomic explosion in the atmosphere corresponds to the solution of high nonlinear heat conductivity where the front, surrounding the object, exists for a sufficiently long time. This analogy has helped me to find the explanation of a tornado by the special dependence of turbulent viscosities on the turbulent eddy spin rate.

The eddy spin (surplus angular velocity) is an invariant variable that can serve as the turbulent state parameter. The turbulent flows are thermodynamically open systems in the Schroedinger sense. On other hand, the necessary flux of energy (of "negentropy") is exactly the same as the Richardson - Kolmogorov concept of energy flux along the eddy hierarchy in atmospheric turbulence. In stationary cases this flux is equivalent to energy dissipation, that is, the concept is in accordance with the Prandtl nonlinearity for the Boussinesq viscosity. The enstrophy that was studied by J. Charney and others is also to the point. The reader will understand that such coincidence is not merely occasional.

---

*V.N. Nikolaevskiy. Geomechanics and Fluidodynamics. Dordrecht: Kluwer (1996).

The competition between the Prandtl and Taylor approaches to turbulence: is well known: which diffusion - impulse or vortex - is predominant? The independent angular momentum balance and the spin appearance resolve the dilemma. Some mishmash of the concepts of homogeneity and reflection invariance has to be overcome to permit adequate study of vortex anisotropy, necessary for some astrophysics problems. The restrictions for stress tensor symmetry due to commutability and limit reasons are also estimated and eliminated.

The Cosserat continuum was developed further by the efforts of C. Truesdell, J. Ericksen, E. Aero, A. Eringen et al, and a number of effective mathematical models of polar / micropolar / micromorphic / oriented fluids and asymmetric hydrodynamics have been published. The theory was applied to liquid crystals and a recognized success was achieved. These results helped essentially to develop the suggested turbulence theory to its present state.

All these aspects are considered in this book and invaluable inputs of my colleagues - Djavid Iskenderov (Kiev), Vladimir Ivchenko (Leningrad), Vladimir Babkin (Petrozavodsk), Oleg Dinariev and Sergey Arsenyev (Moscow), - provided the necessary spectrum of problems under consideration.

Eugenia Kalnay, professor at the University of Maryland, kindly invited me to Washington D. C., where I had the opportunity to become acquainted with up-to-date achievements in the atmospheric sciences.

My talk with G.S. Golitzin was encouraging and useful.

Dr. Edwin Beschler and Dr. Alexander Gubar' helped me with the editing.

The author is indebted to persons mentioned here for their interest and support.

May 2003, Moscow Victor N. Nikolaevskiy

All real fluid motions are rotational. Even in nearly irrotational flows the relatively small amount of vorticity present may be of central importance in determining major flow characteristics, and even some of those whose interest in fluid dynamics is only of the practical sort are now beginning to learn that the hitherto largely neglected question of vorticity must at last be faced.

Clifford Truesdell, 1953

INTRODUCTION

Turbulent flows are full of vortexes, the chaotic motion of which possesses a great amount of kinetic energy, mainly in the form of angular rotation. The question is how to find the special features of this state and use them in continuum balances. In engineering the only way is practical, that is, physically measuring forces and calculating the corresponding velocity field. Forces are measured at a plane of gage (across the tube flow section, etc) while energy and impulses are contained in a fluid volume. That is why we need to begin with spatial averaging to determine turbulent flow parameters.

The *spatial averaging*, mentioned originally by O. Reynolds for a turbulent flow, includes the forces and impulse fluxes acting at oriented cross-sections. Actually, any stress tensor is clearly the dyad of the force and the normal to the mentioned plane and this corresponds exactly to the Cauchy stress determination. It means *possible asymmetry of the Reynolds turbulent stress tensor* and needs a nontrivial angular momentum balance. The balance has to be added to the conventional dynamic equations that are incomplete due to internal bulk couple forces.

The *angular balance* with the turbulent sub-scale eddy spin velocity is added to mass and impulse balances. This corresponds to a new chain of moments of growing order that are found by multiplication of the Navier - Stokes equations by a radius-vector relative to the mass-center of a coordinate cell, and their consequent averaging. The comparison of angular momentum with the vortex representation permits us to find the *inertia tensor evolution* for the turbulent eddy. The latter coincides with the equation for a square of length scale of a turbulent *mesovortex*.

The result is the *Cosserat theory, uniting* the classical theory of turbulence in the *Prandtl-Taylor-Karman formulation with the pioneering ideas of Mattiolli.*

The so-called "semi-empirical" kinetic laws for turbulent viscosity correspond to the representation of turbulent flows in an open thermodynamics system governed by the Kolmogorov flux of energy. The historical aspects of the angular momentum concept development are taken into consideration.

The *spatial averaging* is the only one that *corresponds to the Newton principle* of continuum mechanics when the forces are introduced as the reaction of the ambient space. However, the procedure, which is conventional for turbulence, is based on averaging in time and on the measurement of velocity changes by thermocouples in a single point or pair of points. Therefore, the approach of the present study has to be motivated by the underlying details. The *averaging has to be performed over a coordinate cell* and the Ostrogradsky-Gauss theorem simplifies the situation because of smoothness of velocity fields due to true fluid viscosity. The multi-scale hierarchy of turbulent structure makes the procedure non-trivial.

Turbulent flows belong to *open thermodynamics systems* because of the energy flux from an average flow to turbulent eddies level(s) and then to molecular chaos. Correspondingly, there are a number of internal energies (for different modes of motion) and there takes place an energy exchange between them. Therefore, the *flux of energy plays the role of a thermodynamic parameter*. This means, for example, that turbulent viscosities are functions of the strain rate or/and turbulent eddy spin invariant to coordinate system changes (but not the strain itself as in the conventional thermodynamics of mechanics of gases or solids).

The half-empirical relations that complete the system are, in essence, constitutive rheologic laws, which are usual for continuum mechanics. Of course, the Prandtl scheme of turbulent eddies mixing inside flow layers is developed to account for the eddies' own (spin) rotation.

However, in turbulence an *internal (mesovortex) structure* is generated by boundaries and, therefore, these laws have to be determined really empirically. Correspondingly, only general features of the constitutive laws can be established a priori on the base of reasonable kinetic assumptions. In *stationary cases the energy flux, estimated by the dissipation rate* of mechanical energy into heat, appears here as the state parameter.

The problems under consideration differ by orientations of their eddy spin axes. The simpler one is the case of eddy rotation inside the plane of an average flow. The first one is the *Karman vortex street* behind a body. The corresponding boundary condition is formulated according to the Zhukovsky study of trailing vortexes. The body of revolution also generates the turbulent attached jet considered here. The parameters of flow are determined by proper experimental data.

One more geophysical problem, considered in this monograph, is the *World Ocean dynamics parameterization* necessary for numerical calculation of global circulation. The macroscopic but differential grid means inclusion of *synoptic eddies as mesoscale objects* and that is why the conventional equations are incompatible.

The *suspended particles* change the meso-flow of fluid. Their *rotation effect* in the cloud in its laminar flow was considered in a frame of interpenetrating continua existing simultaneously in the same macro-point. The analysis of flow in the vicinity of the particle permits us to find an effective expression for rotational viscosity. The bulk spin effect is balanced by anti-symmetric components of shear forces, the symmetric shear forces being described by the Einstein formulae.

The combination of turbulent eddies and rotating solid particles can also be approached with the same tools and despite a formally complicated set of dynamic equations permits us to show the *stabilizing effects of the suspension on the Karman street flow*. This problem can give useful estimations for the ecological influence on the atmosphere of the chimney ash and dust.

Rotational inertial effects cause much longer dusty wakes in the atmosphere as well as *the stabilization* of turbulent objects such as tornadoes or hurricanes. *The tornado* is one of the most intriguing geophysical phenomena and its description is based on a nonlinear variant of the theory under consideration.

Possible dependence of turbulent viscosities on spin eddy rotation rate leads to nonlinear diffusion equations. This means the localization of a turbulent object by a front slowly spreading in the atmosphere. *The mathematical problem of the tornado is somewhat similar to the "fire ball" of atomic explosion in the atmosphere*. Two variants are developed: with constant inertia moment and with its evolution. The latter gives the solution with internal core ("eye") rotating in a solid manner. The coincidence with hurricane David is shown as a case study.

One more case that can be treated by the idea of *interpenetrating continua* is that of *turbulent flow with laminar spots*. The latter can be considered as a suspension of fluid particles with the same density but in a different state. The coincidence with the experimental data underlines the effective diffusion mechanism for spreading of laminar spots.

Circulation in the World Ocean can be numerically computed, but the large linear scale of the *effective grid is much bigger than the synoptic eddies* radius. This makes the situation analogous to the plane turbulence but on the rotating

Earth surface. The solution gives a more detailed Globe circulation description in comparison to conventional approaches, with rates closer to experimental data and with more distinct local circulation centers.

The equatorial counter-flows are also explained by the angular momentum effects. The *Cromwell and Lomonosov equatorial flows* in the Ocean had not previously been adequately described due to their orientation change along the vertical axis. The explanation was found in their asymmetrical turbulence and is directly connected with the effect of angular momentum of eddies generated at the bottom and by the wind at the Ocean surface.

Magnetic turbulence as well as differential rotation *demands the averaging along a line* segment for a proper account of electromagnetic fields. The asymmetric turbulence of magnetic fluids permits us to find new dynamo effects in electro-magnetic flows and this can be essential for some potential applications.

The conventional statistical theory, based on averaging over an ensemble of random field realizations, had practically no room for *anisotropy of vortex type* in the homogeneous case. The author has found that the corresponding *prohibitive statement had mixed up the homogeneity and mirror reflection properties*. If one separates these requirements, the study of the vortex anisotropy becomes possible.

The turbulence generated by a wall's asperity is characterized by a change of eddy axis by its evolution in space. At the very beginning the spin axis is orthogonal to a mean flow and parallel to a wall itself. Then it may conserve its position (the Couette flow) or be reoriented by a mean flow by action of another wall (the channel flow). In this case the simple scheme of the turbulent Prandtl-Karman viscosity can be replaced by *the vector-director effect as in the theory of liquid crystals* of Ericksen-Leslie.

Turbulence theory, as evidenced by the above discussion, is clearly of crucial interest to a wide variety of students, engineers and research specialists in natural sciences, hydrodynamics, geophysics, continuum mechanics, ecology, atmospheric and oceanic studies, hydraulics, and other related fields. We hope this monographs will provide some insight and direction in these fascinating studies.

CHAPTER 1

ANGULAR MOMENTUM IN A VISCOUS FLUID

1.1. DYNAMICS OF VISCOUS FLUIDS

The initial differential equations valid for any fluid flow are the balances of mass, impulse and energy for differential volume dx^3:

$$\frac{\partial \rho}{\partial t} + \frac{\partial \rho\, u_j}{\partial\, x_j} = 0, \tag{1.1.1}$$

$$\frac{\partial \rho u_i}{\partial t} + \frac{\partial \rho u_i u_j}{\partial x_j} = \frac{\partial \sigma_{ij}}{\partial x_j} + F_i, \tag{1.1.2}$$

$$\frac{\partial}{\partial t}\,\rho(\mathrm{E} + \frac{u_i\, u_i}{2}) + \frac{\partial}{\partial\, x_j}\,\rho(\mathrm{E} + \frac{u_i\, u_j}{2})\, u_j = \frac{\partial\, \sigma_{ij}\, u_j}{\partial\, x_j} + F_i\, u_i - \frac{\partial\, q_j}{\partial\, x_j} + Q, \tag{1.1.3}$$

Here ρ is a density, u_i is a local velocity, σ_{ij} is a stress tensor, F_i is a mass force, E is a specific internal energy, q is a heat flow, Q is a bulk heat source.

System (1.1.1) - (1.1.3) has to be supplemented with the angular momentum balance equation. Conventionally in mechanics [287] this equation is expressed as equality of tangential stresses and is used in the model of a viscous fluid:

$$\sigma_{ij} = \sigma_{ji}. \tag{1.1.4}$$

The stress tensor limited by condition (1.1.4) is symmetrical. Later we shall consider cases when that *symmetry condition* (1.1.4) *can be violated due to the turbulence effect*. This is the key to the theory developed in this book and it will be valid in a coordinate system with grid scale bigger than eddy sizes.

Multiply impulse equation (1.1.2) by velocity u_j. Due to the following equalities:

$$u_i(\frac{\partial \rho\, u_i}{\partial t} + \frac{\partial \rho\, u_i\, u_j}{\partial x_j}) = \rho\,[(\frac{\partial}{\partial t} + u_j \frac{\partial}{\partial x_j})\, \frac{u_i\, u_i}{2}] \equiv \rho\, \frac{d}{dt}\, \frac{u_i\, u_i}{2};$$

$$u_i\, \frac{\partial\, \sigma_{ij}}{\partial\, x_j} = \frac{\partial\, u_i \sigma_{ij}}{\partial\, x_j} - \sigma_{ij}\, \frac{\partial u_i}{\partial\, x_j}$$

the product is evidently the balance of kinetic energy:

$$\rho \frac{d}{dt}\frac{u_i u_i}{2} = \frac{\partial\, u_i \sigma_{ij}}{\partial\, x_j} - \sigma_{ij}\, e_{ij} + F_i u_i \tag{1.1.5}$$

Here the strain rate e_{ij} appears as a symmetrical part [137, 169] of the distortion rate tensor d_{ij}:

$$e_{ij} = \frac{1}{2}(\frac{\partial\, u_i}{\partial\, x_j} + \frac{\partial\, u_j}{\partial\, x_i}); \qquad d_{ij} = \frac{\partial\, u_i}{\partial\, x_j}$$

The work done against rotation with the angular velocity Φ_k about the center of differential volume element dx^3

$$2\Phi_k = \varepsilon_{kij}\, \frac{\partial\, u_i}{\partial\, x_j} \equiv 2\varepsilon_{kij}\Phi_{ij}$$

is zero because equality (1.1.4) neglects antisymmetric stresses. Here $2\Phi_k$ is the vorticity pseudo-vector, ε_{kij} is the Levi-Civita alternator.

Subtracting (1.1.5) from (1.1.3) gives the heat balance (the first thermodynamics law):

$$\rho\, \frac{dE}{dt} = -\, p\, e_{ij}\delta_{ij} + \sigma'_{ij}\, e_{ij} - \frac{\partial\, q_j}{\partial\, x_j} + Q, \tag{1.1.6}$$

The stress deviator $\sigma'_{ij} = \sigma_{ij} + p\, \delta_{ij}$ and pressure $p = -\,\sigma_{ij}\, \delta_{ij} / 3$ appear here separately. The substantial derivative

$$d / dt = \partial / \partial t + u_j \partial / \partial x_j$$

as well as the unit Kronecker tensor δ_{ij} is used also.

The entropy s of the fluid is introduced altogether with temperature T:

$$\rho \frac{dE}{dt} = - p\, e_{ij} \delta_{ij} + \rho T \frac{ds}{dt} \tag{1.1.7}$$

Equality (1.1.7) is called the Gibbs relation revealing the parameters of the thermodynamic state. In our case the state equation has the form $\rho = \rho\, (T,\, p)$, because

$$\rho\, d(1/\rho)/dt = e_{ij}\, \delta_{ij}.$$

Introduce the dissipation $D = \sigma^{\prime}_{ij} e_{ij}$ as a source of entropy growth corresponding to the viscous model:

$$\rho T \frac{ds}{dt} \equiv \sigma^{\prime}_{ij}\, e_{ij} - \frac{\partial\, q_j}{\partial\, x_j} + Q \tag{1.1.8}$$

Equation (1.1.8) may be transformed to the obvious form of the entropy balance equation on the scale $l \sim dx$:

$$\frac{\partial}{\partial t} \rho\, s + \frac{\partial}{\partial\, x_j} \rho\, s u_j = \Xi + \frac{\partial}{\partial\, x_j} \frac{q_j}{T} + Q \tag{1.1.9}$$

Here q_j/T is the flux of entropy, Ξ is local entropy production:

$$\Xi = \frac{1}{T} (\sigma_{ij} \frac{\partial u_i}{\partial x_j} - \frac{q_j}{T} \frac{\partial T}{\partial\, x_j})$$

We neglect the heat source Q. Then one finds that the entropy production has the form:

$$\Xi = \frac{1}{T} \sigma^{\prime}_{ij}\, e_{ij} - \frac{q_j}{T^2} \frac{\partial T}{\partial\, x_j} \equiv \sum_n X_n Y_n \tag{1.1.10}$$

According to the Onsager principle [63] the thermodynamic forces X_n and fluxes Y_n in the first approximation are proportional to each other, that is,

$$X_n = \sum_m A_{nm} Y_m$$

In addition we take into account the Pierre Curie principle of coincidence of tensor ranks of corresponding fluxes and forces, that is, the stresses in a viscous fluid are proportional to a strain rate:

$$\sigma'_{ij} \equiv \sigma_{ij} + p\delta_{ij} = \nu_{ijkm} e_{km} \tag{1.1.11}$$

$$\nu_{ijkm} = \rho(\upsilon - \frac{2}{3}\nu)\, \delta_{ij}\, \delta_{km} + 2\rho\, \nu\, \delta_{ij}\, \delta_{km},$$

Here $\rho\, \nu$ is the shear dynamic viscosity, ν is shear kinematical viscosity and $\rho\upsilon$ is bulk dynamical viscosity.

We also obtain a heat flux proportional to the temperature gradient:

$$q_i = -\ \kappa \frac{\partial T}{\partial x_i} \tag{1.1.12}$$

where κ is the coefficient of thermoconductivity.

In the case of fluid incompressibility, substitution of rheological law (1.1.11) into equation (1.1.2) determines the Navier-Stokes equation:

$$\frac{\partial \rho u_i}{\partial t} + \frac{\partial \rho u_i u_j}{\partial x_j} = -\frac{\partial p}{\partial x_i} + \frac{\partial}{\partial x_j}\left(\rho\, \nu \frac{\partial u_i}{\partial x_j}\right) + F_i \tag{1.1.13}$$

This impulse equation can be transformed [69, 137] into the form that is for the vorticity $\Omega_i = 2\ \Phi_i$ (the angular velocity Φ_i):

$$\frac{d\Phi_i}{dt} = \left(\Phi_j \frac{\partial}{\partial x_j}\right) u_i + \nu \frac{\partial^2 \Phi_i}{\partial x_j \partial x_j} \tag{1.1.14}$$

Equation (1.1.14) is the diffusion with convection of a vortex in a plane case:

$$\frac{d\,\Phi}{d\,t} \equiv \frac{\partial\,\Phi}{\partial\,t} + u_i \frac{\partial\,\Phi}{\partial\,x_i} = \nu \frac{\partial^2\,\Phi}{\partial\,x_j\,\partial\,x_j} \tag{1.1.15}$$

In the axially symmetric case this has the well-known Fourier linear form:

$$\frac{\partial\,\Phi}{\partial\,t} = \nu\left(\frac{\partial^2\,\Phi}{\partial\,r^2} + \frac{1}{r}\frac{\partial\,\Phi}{\partial\,r}\right) \tag{1.1.16}$$

$$\Phi = \frac{1}{r}\frac{\partial (r\,u)}{\partial r} \; .$$

Here u is the angular component of fluid velocity.

The formal solution of (1.1.15), describing heat conductivity from an instant source, is the following one:

$$\Phi = \frac{A}{2\,\nu\,t}\exp\left(-\frac{r^2}{4\,\nu\,t}\right), \qquad A = const.$$

$$u = \frac{A}{r}\left[1 - \exp\left(-\frac{r^2}{4\,\nu\,t}\right)\right], \qquad u(0) = 0 \tag{1.1.17}$$

However, kinetic energy, angular momentum and dissipation are represented by divergent integrals [260].

The angular momentum in fluid is defined as [136]:

$$M_i = \int_V \rho\,\varepsilon_{ijk}\,r_j\,u_k\;dV \tag{1.1.18}$$

The following particular value of angular momentum corresponds to the solution, suggested by G. I. Taylor [68, 277] and has been obtained by time differentiation of (1.1.17):

$$\Phi = \frac{\lambda^2}{\nu\,t^2}\left(1 - \frac{r^2}{4\,\nu\,t}\right)\exp\left(-\frac{r^2}{4\,\nu\,t}\right) \tag{1.1.19}$$

$$u = \frac{\lambda^2\,r}{2\,\nu\,t^2}\exp\left(-\frac{r^2}{4\,\nu\,t}\right)$$

$$M = 2\pi\,\rho\int_0^\infty r^2 u\;dr = 8\,\pi\,\rho\,\nu\,\lambda^2 \tag{1.1.20}$$

This value M is independent of time and the dimension of constant λ is length. Taylor estimated the effective eddy's radius as $r_* = \sqrt{2\,\nu\,t}$.

Then its moment of inertia and the rotation velocity are as follows:

$$\rho I_* = 2\pi \rho \int_0^{r_*} r^3 dr = \frac{\pi}{2} \rho r_*^4 = 2\pi \rho \nu^2 t^2$$

(1.1.21)

$$\frac{1}{2} \Phi_* = \frac{8\lambda^2}{r_*^2} \frac{1}{t} = \frac{4\lambda^2}{\nu t^2}$$

The scaled moment of inertia is growing proportionally to time:

$$I_* /(\pi r^2) = \nu t \qquad (1.1.22)$$

The diffusion redistributes motion only to a bigger scale. The eddy's radius growth is compensated by decrease of its angular velocity in time.

G. Batchelor [21] considered the angular momentum of the fluid spherical element that will be applied below. He defined it as

$$M_i = \int_V \rho \, \varepsilon_{ijk} r_j (u_k + r_l \frac{\partial u_k}{\partial x_l}) \, dV \qquad (1.1.23)$$

and assumed that velocity and its distortion were *constant* in V. (This is equivalent to the rigid-type rotation.) Then the first part of integral (1.1.23) is zero and the second part defines the inertia moment ρI:

$$M_i = \varepsilon_{ijk} \frac{\partial u_k}{\partial x_l} \int_V r_j r_l \rho \, dV = 2\rho I \Phi_i \qquad (1.1.24)$$

Note that equation (1.1.2), multiplied by the x_k coordinate, will allow us to express the *impulse flux*:

$$\sigma_{ik} - \rho u_i u_k = -\frac{\partial \rho u_i x_k}{\partial t} -$$

(1.1.25)

$$\frac{\partial \rho u_i u_j x_k}{\partial x_j} + \frac{\partial \sigma_{ij} x_k}{\partial x_j} + F_i x_k$$

Multiplication of (1.1.25) by the Levi-Civita alternating tensor ε_{lik} transforms this equation into the angular momentum balance for the differential volume $dx_i dx_k dx_j$ of the coordinate system:

$$\frac{\partial}{\partial t}(\varepsilon_{lki}\,\rho\,u_i\,x_k) + \frac{\partial}{\partial x_j}(\varepsilon_{lki}\,\rho\,u_i\,x_k\,u_j) =$$

$$\frac{\partial}{\partial x_j}(\varepsilon_{lki}\,\sigma_{ij}\,x_k) + \varepsilon_{lki}\,F_i\,x_k \qquad (1.1.26)$$

Integrate equations (1.1.1) - (1.1.3) over the volume V. Due to continuity of all variable fields in a viscous fluid, one can apply the Ostrogradskiy-Gauss theorem, that is, transform volume integrals of divergent forms into surface integrals, taken over all surfaces S of the volume V:

$$\frac{\partial}{\partial t}\int_V \rho\,dV + \int_S \rho\,u_i\,dS_j = 0, \qquad (1.1.27)$$

$$\frac{\partial}{\partial t}\int_V \rho\,u_i\,dV + \int_S \rho\,u_i\,u_j\,dS_j = \int_S \sigma_{ij}\,dS_j + \int_V F_i\,dV, \qquad (1.1.28)$$

$$\frac{\partial}{\partial t}\int_V \rho\left(\mathrm{E} + \frac{u_i\,u_i}{2}\right)dV + \int_S \rho\left(\mathrm{E} + \frac{u_i\,u_i}{2}\right)u_j\,dS_j = \qquad (1.1.29)$$

$$= \int_S \sigma_{ij}\,u_i\,dS - \int_S q_j\,dS_j + \int_V Q dV + \int_V F_i\,u_i\,dV$$

The integration of (1.1.26) yields the balance of an angular momentum of a fluid in the finite volume V:

$$\frac{\partial}{\partial t}\int_V \varepsilon_{lki}\,\rho\,u_i\,x_k\,dV + \int_S \varepsilon_{lki}\,\rho\,u_i\,x_k\,u_j\,dS =$$

$$\int_S \varepsilon_{lki}\,\sigma_{ij}\,x_k\,dS + \int_V \varepsilon_{lki}\,F_i\,x_k\,dV \qquad (1.1.30)$$

1.2. ANGULAR MOMENTUM IN HYDRODYNAMICS (REVIEW)

The key point of our study is the use of vorticity as a kinematical parameter, determined by a velocity field, in the angular momentum of a fluid element. The conjugate variable of vorticity is the inertia moment which concept is not clear especially for a fluid continuum. This was a point that has disturbed a number of researchers.

On the other hand, C. Truesdell [286] and L. Howard [112] discussed possible attempts to find divergent fluxes, involving vorticity. The angular momentum balance could be such a representation if the inertia tensor is determined.

N. Kochin [136] discussed the concept of angular momentum including the inertia tensor; he also discussed the rotation in details but with application to a solid body (for the application to a fluid volume, see [257]).

The inertia moment about the rotation axis for a fluid can be found in the early Poincare theory of vortexes [237] where he considered angular momentum of the vortex tubes. H. Poincare found that in the absence of viscous friction, this parameter is constant. However, in the case of viscous diffusion, as he proved, the inertia moment was growing proportionally to time. The Taylor vortex given by (1.1.19) determines such an inertia moment that does grow in this manner.

G. Batchelor [21] has pointed out that the non-divergent term in the three-dimensional vortex equation is connected with changes of the inertia tensor of fluid element, but the corresponding equation for this object's evolution was not written in the cited book. It was found and given in Section 1.3 under consideration of changes of a fluid element filling the coordinate cell.

P. Saffman [257] considered angular momentum of the finite fluid volume separating the inertia tensor, G. Batchelor limited himself [21] to the case of infinite fluid volume and J. Lumley [173] repeated the latter consideration of the angular momentum for a large fluid volume, assuming additionally that there is some asymmetric part of angular momentum, connected with fluid rotation. As it was done in Section 1.2, Lumley used the method of multiplication of the impulse balance by a radius-vector and integrating over volume. However, he thought that all impulse fluxes are symmetric and this assumption excluded further steps in his study. Later, D. Straub and M. Lauster [270] discussed inadequacy of the Navier-Stokes equations to account for angular momentum conservation.

Let us assume now, following D. Condiff and J. Dahler [50] that "a little line of flowing fluid can rotate with its vorticity and meet with a resistance or stress due to the internal spin if vorticity and internal spin are not synchronized". Then this resistance is a function of the difference angular velocities difference. According to [50] this observation was evidently first made by M. Born[1].

In his kinetic theory [90] Y. Frenkel took account of the independent rotation of long molecules in a liquid, introducing the special vortex viscosity (see also [63]). V. Sorokin [267], V. Nemtzov [205] and many others further developed this approach using continuum and statistical methods.

The suspension of rotating solid particles is another good example of internal, kinematically independent rotation. J. Happel and H. Brenner [104] considered the torques acting on rotating spheres and other bodies inside a fluid. If one tries to account for such effects for a flow of suspension, the necessity of introducing stress asymmetry will be obvious for equilibration of the corresponding bulk couples.

E. Afanasiev and V. Nikolaevskiy [3] made this step obtaining the special form of angular balance independent of impulse and mass balances. Besides they applied the scheme of interpenetrating continua known in translation dynamics of gas mixtures (C. Truesdell [288]) and suspensions (H. Rakhmatulin [244]).

D. Condiff and H. Brenner suggested [49] a diffusion process of rotating particles in a fluid flow.

M. Shliomis [264], A. Tsebers [290] A. Listrov [167], R. Hsieh [113] and other authors, who considered the suspension of ferromagnetic particles, suggested very similar mathematical models (see also [207]). Such a suspension can rotate under action of an external magnetic field. It was shown experimentally that the magnetization of the suspension is proportional to an angular momentum [113].

The corresponding type of hydrodynamics was named *asymmetric* (E. Aero et al [1, 2], J. Dahler and L. Scriven [57, 58]) or *orientation* or *(multi) polar* [80, 289]. All of them are a part of the general continuum suggested by the brothers F. and E.

---

[1] Z. Physik 1, 221, 1920.

Cosserat [54] one hundred years ago. They were the first to deduce the balance equation of internal angular momentum of continua.

C. Truesdell [289] and J. Ericksen [77-79] revived the common interest in the Cosserat continuum. After this, a lot of papers were devoted to the corresponding chapter of continuum mechanics. Specialists in mechanics know well the subsequent enthusiasm that occurred everywhere for moment theories, which led to recognized success in the liquid crystals theory [1, 65, 77-81, 160]. E. Aero and A. Bulygin [2] as well as M. Shahinpoor [261] published reviews of corresponding studies.

A. Eringen [80, 81] and some of his colleagues [4, 5] suggested use the *micropolar* fluid model directly *for turbulent flows* considering small scale turbulent eddies as *micro*elements of a flow. Eringen introduced the *microinertia tensor* J_{ij} and the *microrotation* ν_{ij} effect was taken into account in inertia moment evolution.

The main objection to this model was the conventional principle of the commutative property of velocity components in the Reynolds stress expression:

$$R_{ij} = -\overline{\rho\, u_i u_j} \equiv -\overline{\rho\, u_j u_i} = R_{ji} \tag{1.2.1}$$

The assumed tensor symmetry was in agreement with the hot-wire technique of velocity fluctuation measurements [96] and the developed statistical theory of homogeneous turbulence [20].

Therefore another averaging method had to be suggested and this is the *spatial averaging*. This means [212, 302] that a turbulent stress has to be determined as an impulse $\rho\, u_i$ flux through the oriented plane with normal velocity component u_j. (In [302] the consequences analyzed here were missed.)

In his review of the statistical theory of turbulence H. Dryden [67] wrote about velocity components: "In the most cases *U, V,* and *W* are averaged values at a fixed point over a definite period of time, although in certain problems it is more convenient to take averages over a selected area or within a selected volume at a given instant". Only in this case can it be made evident that "each stress component is thus equal to the rate of transfer of momentum across the corresponding surface by the fluctuations."

The spatial averaging procedure was the main point of the special discussion[2]. That is why we should pay great attention to the spatial averaging procedure in

[2] See a special issue "Problems of averaging and of construction of continual models within continuum mechanics" (Moscow State University Press, 1980, 93 pp).

this book. Principles of the theory suggested in this book were developed mainly in the 70 years of the twentieth century [212-223] but many elements of the theory were approved also in studies of other authors.

First of all, O. Reynolds himself kept the tangential turbulent stress components R_{ij} and R_{ji} as different in the original paper [245] and mentioned the importance of moment of momentum for turbulent flow.

In 1933 Italian researcher G. Mattioli [180-182] was the first to consider the *angular momentum of turbulent fluid* evidently as fundamental and independent equation of turbulence. Mattioli had introduced all the necessary additional characteristics of a turbulent state, that is, an eddy as element object of turbulence, the antisymmetric part of the Reynolds stresses, bulk couples and angular momentum that included the inertia moment. The latter is estimated via molecular viscosity and characteristic velocity[3]. However, Mattioli was not acquainted with the pioneering works of the brothers Cosserat on a general continuum and therefore *identified the eddy angular velocity with average field rotation.* That is why Mattioli used the independent angular momentum balance directly for determination of the turbulent viscosity.

Nevertheless, at the time Mattioli's direction was not accepted and was developed only subsequently[4]. Although in 1934 T. Karman noted "an interesting theory of turbulent transfer", he thought that the Mattioli forces are "some incomprehensible". The Mattioli works were not referenced even in bibliographies in monographs of encyclopedic character.

Actually, C. Ferrari [87] found the assumptions about mesovortex spin to eliminate disagreement of the Mattioli theory with the general three-dimensional model of asymmetrical hydromechanics. Ferrari underlined the necessity to introduce the surplus angular velocity (spin) that appeared in the angular momentum balance. The rotational viscosity was assumed to have a molecular origin (with reference to Y. Frenkel [90]). However, Ferrari assumed that spin is proportional to the second derivative of translation velocity in respect to a coordinate. Therefore the spin velocity lost its kinematical independency, but only this assumption permits us to get the two Mattioli equations: the first for the velocity profile (at a wall it was exactly logarithmic distribution) and the second one for the turbulent viscosity profile across the flow. Besides, turbulent flows at

[3] It was very close to the interpretation of the inertia moment evolution given here.
[4] Gian Domenico Mattioli was tragically murdered in 1946, see [87].

a half-infinite plate, in a wake (jet) and in a curved channel (see experiments by F. Wattendorf [302]) were considered improvements on the Mattioli results[5].

The good agreement with experiments was announced.

At one point A. Kolmogorov [141] suggested three turbulent dynamical equations: for translation velocity, for energy and for some frequency. That is, Kolmogorov approached intuitively the Mattioli system. In his comments to "A.N. Kolmogorov. Selected Works." (Moscow, Nauka, 1985), A. Yaglom noted that the corresponding equation is for the internal length scale of turbulence.

In our works [212, 213], as it will be described in detail below, the spin velocity is found jointly with the average velocity field by solution of the system of balances in combination with the assumed constitutive laws. The latter is formulated in accordance with thermodynamic principles for open system, as it will be given in Chapter 3. That is, the Kolmogorov flux [138-140] of turbulent energy, or the rate of dissipation, plays the role of the constitutive parameter [218, 222]. The suggestion of J. Smagorinsky [266] and V. Novozhilov [227, 228] to use the turbulent viscosity, depending on the second invariant of strain rate tensor, supports this conclusion. J. Ferziger [88], in his turn, used two expressions for turbulent viscosities – average-square values of the strain rate and of the vorticity field. The results were practically identical.

One could see that *the enstrophy concept*, used by J. Charney [43] et al in atmospheric science, is equivalent to energy flux because it is a square root of the scalar product of vorticity. That is why the enstrophy flux was compared successfully with the Kolmogorov scaling of eddy cascade.

M. Lurie and N. Dmitriev [175] used the combination of turbulent energy and the flux of energy to treat the data of the known E. Comte - Bello [48] and J. Laufer [155, 156] experiments. The reason, why they obtained a good correlation, will be discussed also in Chapter 3.

If one considers the simplest micropolar theory, then a condition $\omega = 0$ should be satisfied at the axis of the flow because of symmetry, i.e. spin velocity itself couldn't create the turbulent anisotropy. The corresponding discussion will

---

[5] Attempts to apply this theory to specific problems were undertaken by H. Gurgienko [100 – 102]. It was L. G. Loitzanskiy (who wrote the introduction for the Gurgienko article) and G. I. Petrov, who turned my attention to these works. Petrov had noted an arbitrariness of the inertia moment choice in Mattioli's theory. My point reflected in this book is that all flow parameters have to correspond to a volume element of the selected coordinate system.

be found in Chapter 7 of this book. Of course, the difficulties can be resolved by available experimental fine structure of turbulent flow nearby the wall.

The stereoscopic observations of A. Praturi and R. Brodkey [241] have a special importance in this respect. It follows from their data that transversal eddies arise at the wall due to instability of the laminar sublayer. Then eddies flow away from the wall, their sizes grow, eddies weaken and turn around.

The boundary conditions at a wall have to be formulated for this process. The problem of the adequacy was discussed in many treatises on asymmetrical mechanics. We cite here the books [186], [235] where a number of solutions were given.

Interaction of turbulent parameter fields takes place because of nonlinearity of equations. U. Frost and J. Bitte [92], see also [24], used the Tennekes - Lumley statement that can be interpreted as the energy flux between levels of motion:

$$\frac{d\Phi_i}{dt} = \Phi_j e_{ij}, \tag{1.2.2}$$

We note also the works by H. Naue [197, 198], who suggested the whole chain of phenomenological equations of growing rank for turbulent flows. The latter can be obtained also by the spatial averaging of the Navier - Stokes equations, see Chapter 2.

Results of space averaging depends on its scale because a sub-scale size of energy contained eddies is changing. So, W. Dryden [69] attempted to study scale effect of averaging directly by comparison of stereoscopic photos of colored turbulent patterns in air. His conclusion was that "for increasing scale of averaging, there is a percentage decrease of main-motion kinetic energy and a percentage increase of eddy kinetic energy".

E. Lorenz [171] mentioned the possibility to describe the multi-scale turbulent motion. A. Leonard [159], see also [250], considered the dependency of the average velocity and stresses on scale of space averaging. To account for the large scale eddies on the impulse balance, he suggested a new determination of the Reynolds stresses:

$$R_{ij} = - < \rho u_i u_j > + \gamma < \frac{\partial u_i}{\partial x_k} \frac{\partial u_j}{\partial x_k} > \tag{1.2.3}$$

The discussion of calculation grid choice and connected parameterization of turbulent field can be found in the book edited by F. Nieuwstadt and H. Van Dop [208] and in the review by R. Rogallo and P. Moin [250]. J. Ferziger [88] suggested the two-scale spatial averaging in the practice of a calculation grid selection.

Then $< \overline{u}_i \overline{u}_j > \neq \overline{u}_i \overline{u}_j$. Here both symbols correspond to spatial averaging but bar (over a variable) belongs to a smaller scale of averaging. However, these authors missed the effect of a possible stress asymmetry interconnected with spatial averaging.

Y. Nemirovsky and J. Heinloo [106, 200-203] as well as Y. Berezin and V. Trofimov [27-29] tried to avoid spatial averaging absolutely, introducing their own motivation for the Reynolds stress asymmetry in a number of papers. Their "local – vortex" or "non-equilibrium" models of turbulence seem to be practically equivalent to the theory [212-215], suggested earlier and considered here, and therefore many original solutions presented in their publications have to be mentioned. Between them, there are turbulent flows in external electromagnetic fields [200], the generation of spin by electric fields in [200]; rotational flows in tubes [202]; the convection generated by density pulsations due to temperature spreading [27]; the special evolution of the velocity-temperature field in rotating tubes; heat exchange in the field of turbulent oriented flow [283-285, 303], etc.

A number of corresponding solutions can be found also in papers on micropolar hydrodynamics [6, 128, 132-135, 183, 184]: the Couette flow between concentric cylinders; flow between parallel plates; flow through a cylindrical tube; bulk flow of the flow generated by torque applied to polar molecules by an electric field, etc.

Y. Berezin et al [28] touched the link with the *helicity* concept: "When helicity is non-zero, the flow does not possess reflection symmetry. The Reynolds stresses corresponding to such flows are not symmetric". This opinion contradicts the Moffat concept [188-190]. H. Moffat considered the flows with helicity (that is, with pseudo-scalar product of velocity and vorticity) *without* the stress *asymmetry.*

The symmetry problem can be treated by statistical methods and is represented in Chapter 6. The very important corresponding analysis is identification of the homogeneity concept and mirror-symmetry property. This is essential for astrophysical and earth magneto-dynamics phenomena (F. Busse [38], H. Moffat [190], S. Vainstein, Y. Zeldovich, A. Ruzmaikin [292]).

V. Kruka, S. Eskinazi [82, 83, 148], J. Hinze [110, 111] et al. tried to combine the engineering calculations of turbulent flows with "statistically" determined stresses. They found that extreme points of velocity profiles do not coincide with zero tangent stresses or that negative viscosity had to be used, see also the paper by G. Rudiger [253] and the book by V. Starr [268].

Spatial averaging can be applied to a heterogeneous continuum, see the book on a porous saturated medium by V. Nikolaevskiy et al [225]. There *the concept of elementary volume* was introduced that is recognized now in the literature on porous media [23]. It was applied later for solid heterogeneous media (composites) where stress asymmetry was revealed [161] and later even in the exotic case of ice movement on an oceanic surface [107, 124] (ice-floes can rotate and keep their kinetic movement during a finite time interval). Here the diffusion process aspects become nontrivial. So, S. Corrsin [52] suggested use of an asymmetric tensor of turbulent diffusion.

In Chapter 4 the asymmetrical mechanics of two-phase mixture in laminar and turbulent flows will be considered. The problem of interaction of a turbulent eddy and a suspended particle was studied theoretically (A. Kolmogorov [142], etc) and experimentally (J. Daily and P. Roberts [59, 60]). Large particles tend to increase turbulence but small ones intensify dissipation. Laminar spots in turbulent flows [66, 145] and turbulent lenses [158, 185] in the ocean or atmosphere can be classified as heterogeneous continua and they are discussed also in Chapter 4.

Although the known physicists I. Tamm [274], De Groot [64] and H. Ukawa recognized possible asymmetry of the stress tensor, in the theory of liquid helium [153], that is one more example of a two-phase continuum that possesses additional rotational degree of freedom (in a form of quantized circulations) [13, 242, 280], only in papers [12, 109] were the Cosserat methods used.

Now let discuss the most intriguing problems of *tornadoes and hurricanes*. The origin of these objects was recognized as a synergetic [103] problem (K. Emanuel [75], E. Lupyan et al [174], M. Kurgansky [149, 150], etc).

However, the mechanical concepts of angular momentum or helicity are unavoidable. K. Emanuel [75] used the conservation of absolute angular momentum about the storm center. M. Lavrentiev and B. Shabat [157] discussed the problem and named the angular momentum as "vortex impulse", see also [257]. V. Lugovzev [172] applies the condition of constant vortex impulse (angular momentum) during evolution of vortex but without its attendant changes that would include bulk turbulent couple forces as well as diffusion of angular momentum.

J. Molinari and D. Vollaro [192], trying to obtain "a clearer picture of the relationship between the *eddy momentum source* and outflow at inner radii, computed the tangential velocity eddy flux convergence". A high correlation was found between azimuth fluxes of angular momentum created by eddies at large radii and central pressure changes in the storm (hurricane Elena). E. Palmen and H. Riehl [230] studied the angular momentum balance for tropical cyclones and estimated the surface tangent forces using the conventional symmetric stress concept. G. Chimonas and H. Hauser [47] considered the flux of angular momentum from the center of the vortex to far zones of the atmosphere in a form of *gravity swirl waves.* This happens with hurricanes but not tornadoes because the latter possess too short rotation periods.

The angular momentum field is important because "in classical two-dimensional inertial stability problems, if the angular momentum decreases with the radius in an axisymmetric vortex, it is inertial unstable."(R. Pielke et al [236]. J. Kossin et al [143] discussed the hurricane's "eye" and its stability due to the effect of *differential rotation.*

D. Lilly demonstrated [164] that *helical flow* was closely associated with rotational characteristic of a super-cell thunderstorm. M. Kurgansky [149, 150] developed the helicity evolution equation adding the moist mass transfer. L. Rothfusz and D. Lilly [252] simulate experimentally the tornado - like vortex, produced by vertical transport of angular momentum generated by helical secondary flow. They concluded on the base of the experiment: "The vortex derives its rotation by accumulating the angular momentum from one side of the inflow while rejecting that from the other side. The helicity of the input flow conserved and amplified by inflow acceleration, is responsible for the necessary vertical flux divergence of angular momentum". These authors used the following expression for the angular momentum *valid for rigid - like rotation* with angular velocity Φ:

$$\frac{\Gamma}{2\pi} = u_\theta r = \Phi r^2 = J\Phi \qquad (1.2.4)$$

The *circulation* Γ is the most influential parameter controlling the core diameter of a tornado (D. Lilly [164]). D. Lewellen et al [162] studied the internal structure of the tornado by contours of angular momentum $\Gamma/(2\pi)$ of the flow about the center. It was recognized that "the strong cyclonic tangential circulation is produced by the inward advection of air with large absolute angular momentum from tropical cyclone environment" [236]. Some tornadoes include multiple vortices and well-defined rotating air columns within the larger tornado are called "suction vortexes", see the review by R. Pielke et al [236].

The inertia moment for fluid elements rotating in a finite chamber space appeared in the book by E. Gledzer et al [94]. S. Meachem et al [185], considering the resonance instabilities of "meddies" (small lens-like patches), introduced high – order moments of the anomaly.

J. Egger [71] introduced the angular momentum into global calculations, taking into account of the real spherical geometry of the Earth and into the models f - and β - planes for hurricanes. The conclusion was that the important features were not missed and possible deviations were found. Egger and K. Hoinka [72] performed large-scale calculations of the angular momentum balance in the rotating Earth atmosphere to find out the effect of torques generated by the Earth topographic "asperity".

Most studies of the tornadoes and hurricanes were done without the accent on turbulent stresses. R. Pauley [231] and M. Montgomery et al [196] underlined the role of meso-vortices for axial pressures in tornadoes and hurricanes.

Much more attention was paid to turbulent stresses in astrophysics. In his book J.-L. Tassoul [276] states that the model of a star corresponds to electromagnetic hydrodynamics; all introduced tensors are symmetric; angular momentum is obtained by multiplication of impulse by distance from the star center and integration over the star total volume; the star inertia moment is specially introduced; the transition from differential rotation to solid body is connected with star evolution.

P.A. Gilman published, see [99], a detailed review of the effect of stellar turbulence. He noted the following features: the differential rotation can be expected to be present over a large fraction of the star's life time; its dynamo action creates a magnetic field; rotation affects convection; the latter together with the Coriolis force creates turbulent Reynolds stresses in a stellar gas. The redistribution of angular momentum in a convection zone is potentially important for determining the differential rotation profile. The star would move towards a state of constant angular momentum.

The spiral galactic structure was considered on the basis of classical hydrodynamics (C. C. Lin and F. Shu [166]). The Reynolds stresses were introduced by space averaging and the problem of completing of chain of moment equations appeared. Although the averaging over oriented planes was mentioned, the possible asymmetry of stresses was not discussed at all: the stress tensor is assumed to be symmetric. Under consideration of the rotating Galactic disk, the angular momentum was introduced. The density wave theory of a spiral structure was considered also in the frame of a symmetric stress concept.

The Great Red Spot in the atmosphere of Jupiter is also an interesting problem of turbulence. It is in dynamical equilibrium due to absorbing the small vortices to compensate for dissipation of large vortices [174, 236, 268] (compare with [26]). N. Yanenko et al [300] had formulated the idea that interaction energy between eddies could grow at the expense of their individual energy and this would be the passing of energy from smaller vortexes to bigger ones. If we compare this effect with the Richardson-Kolmogorov cascade of eddies [68, 138, 195, 247], the direction of energy flux down to smaller ones is argued. A number of papers were devoted to this phenomenon.

The problem of turbulence origin is not discussed in this book. Let me mention only that the conventional way consists in the study of stability loss of a laminar flow [126] and in the analysis of growing chaotic motion components [103, 152, 153, 170], compare [26, 224, 297].

1.3. ANGULAR MOMENTUM OF A FLUID ELEMENT

Consider a volume V fixed in some inertia coordinate system and bounded by some surface S. If fields are smooth, we can permits to transform surface integrals to volume integrals. Then balance (1.1.30) of angular momentum relative to a fixed point O can be rewritten for a *multipolar* fluid [210] as follows:

$$\int_V \left(\frac{\partial M_i^*}{\partial t} + \frac{\partial M_i^* u_p}{\partial \xi_p}\right) dV + \int_V \rho\, \varepsilon_{ijk} \left(\frac{\partial r_j u_k}{\partial t} + \frac{\partial r_j u_k u_p}{\partial \xi_p}\right) dV =$$

$$= \int_V \varepsilon_{ijk} \left(\frac{\partial r_j \sigma_{lk}}{\partial \xi_l} + r_j F_k\right) dV + \int_V \left(\frac{\partial m_{ip}}{\partial \xi_p} + C_i^*\right) dV \tag{1.3.1}$$

Multipolar fluids possess the internal angular momentum density M_i^* [57, 58, 125, 269]. Here $r_p = r_p^0 + \xi_p$ is the radius vector originated at O, r_p^0 is the radius vector of mass center of fluid in V, m_{ij} is the couple stress tensor, C_i is the body distributed couples, F_i is the body force.

Assuming that fluid is incompressible ($\partial u_i / \partial \xi_i = 0$) due to the condition $r_p^0 (x) = const$ we can transform (1.3.1) further as:

$$\int_V \left(\frac{\partial M_i^*}{\partial t} + \frac{\partial M_i^* u_p}{\partial \xi_p}\right) dV + \rho\, \varepsilon_{ijk} \{x_j \int_V \left(\frac{\partial u_k}{\partial t} + \frac{\partial u_k u_p}{\partial \xi_p}\right) dV +$$

$$\int_V \xi_j \left(\frac{\partial u_k}{\partial t} + \frac{\partial u_k u_p}{\partial \xi_p}\right) dV + \int_V u_k u_p \delta_{jp} dV\} =$$

$$\varepsilon_{ijk} \{x_j \int_V \left(\frac{\partial \sigma_{kl}}{\partial \xi_l} + F_k\right) dV + \int_V \xi_j \left(\frac{\partial \sigma_{kl}}{\partial \xi_l} + F_k\right) dV\} + \tag{1.3.2}$$

$$\int_V \left(\frac{\partial m_{ip}}{\partial \xi_p} + C_i^*\right) dV + \int_V \sigma_{kj} dV$$

Assuming that

$$\int_V \xi_j dV = 0$$

let us express the kinematical variables in the form of Taylor series with respect to mass center that is moving with velocity U_i:

$$u_i(x_l + \xi_l) = U_i(x_l) + \frac{\partial U_i(x_l)}{\partial x_m}\xi_m + \frac{1}{2}\frac{\partial^2 U_i(x_l)}{\partial x_m \partial x_n}\xi_m\xi_n + o(l^2) \tag{1.3.3}$$

$$\frac{\partial U_i(x_l + \xi_l)}{\partial x_m} = \frac{\partial U_i(x_l)}{\partial x_m} + \frac{1}{2}\frac{\partial^2 U_i(x_l)}{\partial x_m \partial x_n}\xi_n + o(l)$$

As result, (1.3.2) has the form:

$$\rho\,(\frac{\partial M_i{}^*}{\partial t} + U_p\frac{\partial M_i{}^*}{\partial x_p}) + \frac{1}{2}\rho\, I_{ml}\frac{\partial^2}{\partial x_m \partial x_l}(\frac{\partial M_i{}^*}{\partial t} + U_p\frac{\partial M_i{}^*}{\partial x_p}) +$$

$$\rho\,\varepsilon_{ijk}\{x_j[(\frac{\partial U_k}{\partial t} + U_p\frac{\partial U_k}{\partial x_p}) + \frac{1}{2}I_{ml}\frac{\partial^2}{\partial x_m \partial x_l}(\frac{\partial U_k}{\partial t} + U_p\frac{\partial U_k}{\partial x_p})] +$$

$$I_{jl}\frac{\partial}{\partial x_l}(\frac{\partial U_k}{\partial t} + U_p\frac{\partial U_k}{\partial x_p})\ \} + o(l^2) = \varepsilon_{ijk}\{x_j[(\frac{\partial \sigma_{kl}}{\partial x_p} + F_k) +$$

$$\frac{1}{2}I_{ml}\frac{\partial^2}{\partial x_m \partial x_l}(\frac{\partial \sigma_{kp}}{\partial x_p} + F_k)] + \sigma_{kj} + \frac{1}{2}I_{ml}\frac{\partial^2 \sigma_{kl}}{\partial x_m \partial x_l} +$$

$$I_{jl}\frac{\partial}{\partial x_l}(\frac{\partial \sigma_{kp}}{\partial x_p} + F_k)]\} + (\frac{\partial m_{ip}}{\partial x_p} + C_i{}^*) + \frac{1}{2}I_{ml}\frac{\partial^2}{\partial x_m \partial x_l}(\frac{\partial m_{ip}}{\partial x_p} + C_i{}^*)$$

Here $\rho\, V\, I_{ml}$ is the inertia moment tensor of a fluid in V and:

$$I_{ml} = \frac{1}{V}\int_V \xi_m \xi_l dV \qquad (1.3.4)$$

After some recombination, we can transform (1.3.2) further into the sum of terms of the same order:

$$\rho\,(\frac{\partial M_i{}^*}{\partial t} + U_p \frac{\partial M_i{}^*}{\partial x_p} - G_i) - \varepsilon_{ijk}\sigma_{kj} - \frac{\partial m_{ip}}{\partial x_p} + \varepsilon_{ijk}\{x_j[\rho(\frac{\partial U_k}{\partial t} +$$

$$+ U_p \frac{\partial U_k}{\partial x_p}) - F_k - \frac{\partial \sigma_{lk}}{\partial x_l}]\} + \frac{1}{2} I_{ml} \frac{\partial^2}{\partial x_m \partial x_l}[(\frac{\partial M_i{}^*}{\partial t} +$$

$$U_p \frac{\partial M_i{}^*}{\partial x_p} - C_i{}^*) - \frac{\partial m_{ip}}{\partial x_p} - \varepsilon_{ijk}\sigma_{kj}] + \varepsilon_{ijk} x_j I_{ml} \frac{\partial^2}{\partial x_m \partial x_l} \times$$

$$[\rho(\frac{\partial U_k}{\partial t} + U_p \frac{\partial U_k}{\partial x_p}) - F_k - \frac{\partial \sigma_{kp}}{\partial x_p}] + \varepsilon_{ijk} I_{jl} \frac{\partial}{\partial x_l}[\rho(\frac{\partial U_k}{\partial t} +$$

$$U_p \frac{\partial U_k}{\partial x_p}) - \frac{\partial \sigma_{kp}}{\partial x_p} - F_k] + o(l^2) = 0 \qquad (1.3.5)$$

If we take into account impulse balance (1.1.2), then the equation of internal angular momentum will follow

$$\rho(\frac{\partial M_i{}^*}{\partial t} + U_p \frac{\partial M_i{}^*}{\partial x_p} - C_i{}^*) - \varepsilon_{ijk}\sigma_{kj} - \frac{\partial m_{ip}}{\partial x_p} = 0 \qquad (1.3.6)$$

If we consider only conventional fluids (*not micropolar*), their rheology is such that

$$M_i{}^* = 0, \qquad C_i{}^* = 0, \qquad m_{ij} = 0,$$

and the stress tensor is symmetric $\sigma_{ij} = \sigma_{ji}$.

Let the last equalities be valid. Then we get asymptotically at $l \to 0$:

$$\varepsilon_{ijk} I_{jl} \frac{\partial}{\partial x_l}[\rho(\frac{\partial U_k}{\partial t} + U_p \frac{\partial U_k}{\partial x_p}) - \frac{\partial \sigma_{kp}}{\partial x_3} - F_k] = 0 \qquad (1.3.7)$$

Of course, this equation is a consequence of the impulse balance and for any symmetrical volume when $I_{ij} = I\delta_{ij}$, we get as the result of application of the curl operation to the impulse balance:

$$\varepsilon_{ijk} \frac{\partial}{\partial x_j}[\rho(\frac{\partial U_k}{\partial t} + U_p \frac{\partial U_k}{\partial x_p}) - \frac{\partial \sigma_{kp}}{\partial x_3} - F_k] = 0 \qquad (1.3.8)$$

For incompressible viscous fluid this coincides with the classical equation for vortex (1.1.14).

Consider now the inertia moment $\rho \boldsymbol{i}$ of a fluid element $V(t)$ that arise at some instant within the volume V. The geometrical moment $\boldsymbol{i}$ will be calculated about the mass center of this element:

$$i_{ij} = \int_{V(t)} \xi_i \xi_j dV \qquad (1.3.9)$$

Now let us consider the *substation* derivative of the angular momentum of selected fluid element $V(t) \sim l^3$ that corresponds to the Lagrange approach:

$$M_i = \rho\varepsilon_{ijk} \frac{1}{V(t)} \int_{V(t)} \xi_j u_k dV = \rho\varepsilon_{ijk} i_{jl} \frac{\partial U_k}{\partial x_l} + o(l^2) \qquad (1.3.10)$$

that is, the mass center velocity U_i is introduced:

$$\frac{\partial M_i}{\partial t} + U_p \frac{\partial M_i}{\partial x_p} = \rho\, \varepsilon_{ijk} \int_{V(t)} \xi_j \frac{du_k}{dt} dV \qquad (1.3.11)$$

Expression (1.3.11) can be rewritten with the mass center velocity for small volumes:

$$\frac{\partial M_i}{\partial t} + U_p \frac{\partial M_i}{\partial x_p} = \rho\, \varepsilon_{ijk} i_{jl} \frac{\partial}{\partial x_l}\left(\frac{\partial U_k}{\partial t} + U_p \frac{\partial U_k}{\partial x_p}\right) + o(l^5) \qquad (1.3.12)$$

We can integrate (1.3.12) over time for the considered fluid element because at the left – hand side there is *substation* derivative:

$$\frac{\partial}{\partial t} + U_p \frac{\partial}{\partial x_p} \equiv \frac{d}{dt}$$

of the angular momentum:

$$M_i = \rho\, \varepsilon_{ijk} i_{jl} \frac{\partial U_k}{\partial x_l}$$

and the following initial condition is being valid:

$$i_{ij} = I_{ij} \quad (t=0) \tag{1.3.13}$$

In the symmetrical case ($i_{jl} = i\delta_{lj}$) we get the expression (1.1.24) of the *angular momentum of a rotating fluid sphere*:

$$M_i = \rho\, \varepsilon_{ijk} i \frac{\partial U_k}{\partial x_j} = 2\rho\, i\, \Phi \tag{1.3.14}$$

Now we shall study the *evolution* process of the inertia moment:

$$\frac{di_{ml}}{dt} = \int_{V(t)} \frac{d}{dt} (\xi_j \xi_m) dV$$

Because $d\xi / dt = u_m - U_m + o\,(l)$, we have:

$$\frac{di_{ml}(t)}{dt} = \int_{V(t)} (u_m - U_m)\xi_l dV + \int_{V(t)} (u_l - U_l)\xi_m dV + o(l^5) \tag{1.3.15}$$

Following the derivation of (1.3.3), we expand the integrands into series and the limit passing at $l \to 0$ yields:

$$\frac{di_{jl}}{dt} = i_{il} \frac{\partial U_j}{\partial x_i} + i_{ij} \frac{\partial U_l}{\partial x_i} \tag{1.3.16}$$

If at the given instant the fluid particle is symmetrical, then

$$\frac{di_{jl}}{dt} = I \left(\frac{\partial U_j}{\partial x_l} + \frac{\partial U_l}{\partial x_j}\right) = 2\, I\, e_{jl} \tag{1.3.17}$$

We obtain [210] that instantaneous change of the inertia moment is caused only by deformation that may subsequently violate initial symmetry.

Direct differentiation of (1.3.10) yields

$$\frac{dM_i}{dt} = \varepsilon_{ijk} \frac{\partial U_k}{\partial x_l} \frac{di_{jl}}{dt} + \varepsilon_{ijk} i_{jl} \frac{\partial}{\partial x_l} \left(\frac{\partial U_k}{\partial t} + U_p \frac{\partial U_k}{\partial x_p} \right) - \varepsilon_{ijk} i_{jl} \frac{\partial U_p}{\partial x_l} \frac{\partial U_k}{\partial x_p} \tag{1.3.18}$$

The comparison of (1.3.18) with (1.3.12) yields

$$\varepsilon_{ijk} \frac{\partial U_k}{\partial x_l} \frac{di_{jl}}{dt} = \varepsilon_{ijk} i_{jl} \frac{\partial U_p}{\partial x_l} \frac{\partial U_k}{\partial x_p} \tag{1.3.19}$$

The latter can easily be shown also from (1.3.16) and in the case of a symmetrical fluid volume it simplifies to the following one:

$$\varepsilon_{ijk} \frac{\partial U_k}{\partial x_l} \frac{di_{jl}}{dt} = \varepsilon_{ijk} I_{jl} \frac{\partial U_p}{\partial x_l} \frac{\partial U_k}{\partial x_p} = \varepsilon_{ijk} I \frac{\partial U_p}{\partial x_j} \varepsilon_{lpk} \Phi_l \tag{1.3.20}$$

Using the tensor identity:

$$\varepsilon_{ijk} \varepsilon_{lpk} = \delta_{il} \delta_{jp} - \delta_{ip} \delta_{jl} \tag{1.3.21}$$

we finally obtain the following equation [210] for an incompressible fluid:

$$\varepsilon_{ijk} \frac{\partial U_k}{\partial x_j} \frac{di}{dt} = -I \Phi_j \frac{\partial U_i}{\partial x_j} \tag{1.3.22}$$

This means that the first right-hand side term of the Friedman vortex equation for a perfect fluid:

$$\rho \left(\frac{\partial \Phi_k}{\partial t} + U_p \frac{\partial \Phi_k}{\partial x_p} \right) = \rho \Phi_p \frac{\partial U_k}{\partial x_p} + \rho \Phi_k \frac{\partial U_p}{\partial x_p} + \frac{1}{2} \varepsilon_{kip} \frac{\partial F_i}{\partial x_p} \tag{1.3.23}$$

is responsible for the inertia moment change, as it was mentioned also by G. Batchelor [21].

When the volumes of the fluid particles considered are comparable with the cell obtained by sectioning the space by means of the coordinate planes, equation

(1.3.7) and (1.3.11) take the form of the angular momentum balance (in the Euler and Lagrange representation respectively) for differential volume.

In the Cartesian coordinates system the Euler inertia moment of incompressible fluid contained in a cubical cell $V=l^3$ has the form independent of both time and coordinate: $\rho I_{ij} = \rho l^2 \delta_{ij} /12$

In the Lagrange case, the inertia moment i_{ij} is associated with fluid contained in V at the instant t. It changes in time due to rotation and rate.

1.4. MICRO AND MACRO - SCALES

As one can see, two versions of balance equations can be used. The first one is for a small volume element, corresponding to a grid of the selected coordinate system when the balance equations have a differential form. The second one is for large volumes of arbitrary configuration when the balance equations are formulated as integral relations.

This book combines these approaches considering the first elements as micro-volumes and introducing the large volumes in a form of macro - cells, corresponding to a new coordinate grid with much larger linear scale than the initial one. The integral relations are transformed into finite-differential balances absolutely similar to ones used in well - known computer calculations. If these macro-cells themselves are small relative to the external problem scale, the finite-differential relations yield the macro-differential equations including, however, spatially averaged variables. Therefore the key point of the method developed in this book is the spatial averaging procedure.

Such a transition could lead to a trivial result, if the considered medium does not possess a structure that corresponds to an intermediate or *mesoscale* motion. Therefore, the spatial averaging method is effective for a fluid turbulent flow, with eddies having a sub-grid scale. Flows through the porous media, the composite material deforming, and many other phenomena can be treated similarly. Diameter of a pore or a heterogeneous inclusion inside a solid medium, radius of an eddy or a particle suspended in a fluid serves as a mesoscale length. Sometimes, a group of inclusions can play a role of mesostructure and a bigger macro-volume (or *elementary* volume [225]) has to be chosen.

The dynamics of mesoscale objects needs, in corresponding parameters and the spatial averaging method, to introduce naturally additional kinematically independent variables, for example, an *angular velocity* of a suspended particle or *of a turbulent eddy*. Therefore, the adequate total dynamical system has to include more independent equations than in conventional cases. The corresponding way is guided by the Newton mechanics of a finite body. So, the balances of the moments of ranks higher than mass and impulse balances will become non-trivial.

The main advantage of spatial averaging is a possibility to differentiate between the values, averaged at oriented areas, the volume averaged values and even values, averaged along a line. Their non-zero difference appears to be connected exactly with the introduction of additional kinematics degrees of freedom mentioned above and with the necessity to use the dynamics equations of a growing order. The spatial averaging method yields the physical variables and equation systems that were introduced earlier in a phenomenological manner on the basis of some physical reasons. *A posteriori,* comparison of the last ones with the natural data had assured correctness in the developing method.

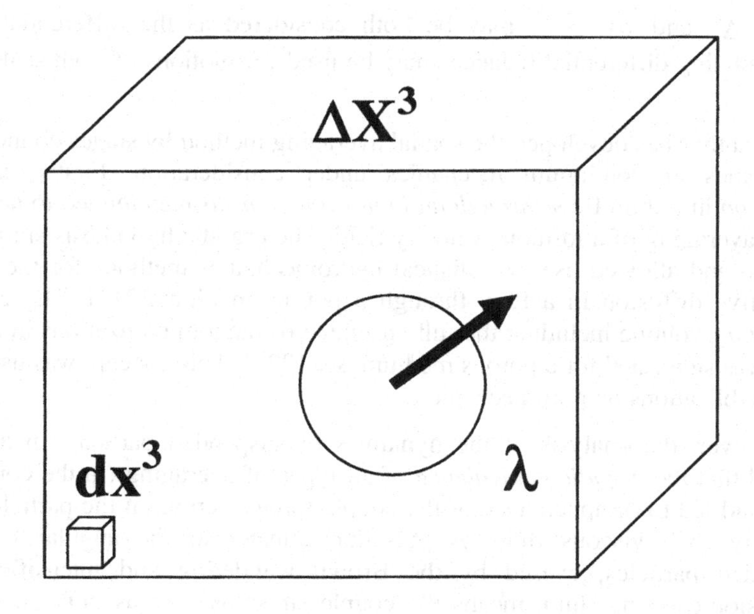

Figure 1.4.1. Coordinate scales to study a turbulent eddy

At the same time, the identical coincidence of differently averaged values leads to splitting of the moment chain and, for example, all balances of the order higher than two are transformed into simple consequences of the first two balances – mass and impulse.

In accordance with these concepts we need to use two linear scales, external L and internal λ, and introduce two coordinate systems, that is, *micro-scale coordinates* x_i ($dx_i \sim l$) and *macro-scale coordinates* X_i ($dX_i \sim \Delta$). These systems subdivide the space into differential volumes $dv = dx^3$ and "elementary" volume $\Delta V = \Delta X^3$, correspondingly.

Let λ - value be a size of turbulent eddy. The L - value is an external scale of a motion (a problem) under consideration. Let us select the scales of coordinates in the following way: $L >> \Delta >> \lambda$, $\lambda >> l >> \lambda_0$, where λ_0 is, for example, molecular ("close to zero") scale. The right-hand sides of inequalities indicate that individual particles, composing continuous variable fields have λ as mesoscale and λ_0 as microscale. Left-hand sides of these inequalities mean that the volumes $\Delta V \sim \Delta^3$ and $dv \sim l^3$ may be both considered as the differential, and the corresponding differential balances may be used for motions of both scales (Δ and l).

The author has developed the spatial averaging method by stages connected with the themes of continuum mechanics under consideration. Firstly, a *tracer's dispersion* in a fluid flow *through an immovable porous medium* led to necessity of spatial averaging of a tortuous velocity field. The ergodic hypothesis simplified the situation and allowed use of statistical hydromechanics methods for the theory of convective diffusion in a flow through a porous medium [211]. The concept of *elementary* volume including the full ensemble of random realizations in a physical space was suggested for a porous medium, see [225]. This concept was used later in many publications on poromechanics.

However, the analysis of the dynamics of suspended particles in a fluid [3] showed that the *ergodic equivalence* of all types of averaging (in the conventional form) had led to disappearance of the couple forces, acting on the particle, rotating relatively to a viscous fluid. A pulsation transfer of the angular moment of suspended particles, caused by the Brown wandering and intensified by the suspension-carrying fluid, means the couple stress as well as bulk couple forces appearance. If one imagines that suspended solid particles instantly transfer into a liquid state, then the picture becomes similar to a turbulent fluid. If just a part of the fluid particles possesses a spin, the picture corresponds to the intermittency of laminar and turbulent spots.

Contrariwise, a "freezing" of dispersing fluid in the suspension leads to a solid composite texture. Certainly, both the kinematics and the force factors are changed under all these "transfers". However, the principles of balance equations averaging must be identical.

CHAPTER 2

SPATIAL AVERAGING AND MACROEQUATIONS

2.1. SPATIAL AVERAGING PRINCIPLES

The averaging problem is one of the central problems in continuum mechanics as well as in physics. In the case of such a complicated system as turbulent fluid, the development of the mathematical model itself depends essentially on averaging.

Spatial averaging (even in non-evident forms) is always used for development of balance field equations in continuum mechanics and, particularly, with the famous Cauchy tetrahedron for illustration of the stress tensor concept.

Averaging methods are contained implicitly in many approximate solutions and, of course, in numerical calculation at large spatial scale or when crossing through the zones of large gradients of variables. Averaging over the cross-section of a flow or of a body is also the conventional procedure that yields simple and efficient engineering calculations, for example, in hydraulics [246].

In the theory of turbulence itself the proper two procedures of introduction of mean flow parameters (for some volume interval and/or for some time interval) were shown even in the Reynolds initial work [245].

The turbulent structure created in a fluid flow by interaction with other objects at high Reynolds numbers can be accounted for on the basis of continuum principles. The most general approach was generated by the Cosserat ideas (see, [54, 125, 269, 289]) and its advantages will be discussed here. The exactly corresponding theory appears after the change to space averaging from the conventional one, used for turbulent flows and assumed to be equivalent to the statistical averaging of an adequate ensemble of realizations.

The spatial averaging procedure corresponds to selection of the coordinate system or discretization grid and leads to the Cauchy determination of the Reynolds turbulent stresses as a force, acting at cell bonding cross-sections. Correspondingly, the stress tensor is introduced as a dyad of the force and the normal at the mentioned oriented planes. Non-equivalency of these two vectors means that their indexes are non-commutable and the tensor is evidently asymmetric. This opens a way to account for the nontrivial angular momentum balance that has to be added to the conventional dynamic equations.

This type of averaging is the only one that corresponds to the Newton principle of continuum mechanics, when the forces are introduced as the reaction of the ambient space to the object under consideration. The averaging has to be performed in a coordinate cell and the Ostrogradsky-Gauss theorem simplifies the situation because of smoothness of velocity fields (due to true fluid viscosity) although the multi-scale hierarchy of turbulent structure makes the procedure non-trivial.

In this connection we would cite the H. Dryden who wrote [67] that in most cases velocities were averaged values

$$U = \bar{u} \tag{2.1.1}$$

at a fixed point over the definite period of time[6], although in certain cases it was more convenient to take averages over a selected area or within a selected volume at a given instant. Each stress component is equal to the rate of transfer of momentum across the corresponding surface by the pulsation velocity (see also, [302]).

The problem of how to choose "*representative*" averaging intervals that include the realizations ensemble is a real one for time averaging [96] as well as for spatial averaging. In the latter case the concept of *elementary* volume was introduced (in 1970 [225]) as a volume that has to be sufficiently small so as to resemble a differential, and simultaneously sufficiently large so as to include the realization ensemble.

Actually, because the considered random velocity field is determined in four-dimensional space (including a time coordinate), the averaging is actually carried out over the generalized space-time volume:

$$\Delta V \times \Delta t$$

Conventional averaging over the time interval Δt at the point x_i (for a particle coinciding with a volume $\Delta V \equiv dv$) is connected with the recognized method of velocity measuring (by thermocouples). The correspondingly developed equations of statistical turbulence theory (Chapter 6) are formed, first of all, for the correlation moments of a velocity field, taken in distant space elements dv and averaged over time. In this procedure it is assumed that the interval Δt is *representative in order* to include the entire ensemble of velocity realizations in the considered elements dv.

---

[6] Time averaging is shown by a bar

The development of mass and momentum equations of, so-called, half-empiric theory for non-stationary (relative to macro-variable quantities $U_i,...$) turbulent flows means actually "instant" averaging over the volume $\Delta V \equiv dV$ or over the cross-section of the flow, which exceeds the corresponding "Prandtl mole" [96, 239, 240] (turbulent sub-scaled eddy) size. The averaging results lead to determined values if the volume ΔV includes the entire ensemble of velocity field realizations. Moreover, the equations formed in this way for turbulent "stationary" fields actually use the averaging over the appropriate (large) volumes ΔV and over some Δt much smaller than intervals ΔT of boundary condition changes.

It is usually thought that the value averaged either over time (Δt) or over volume (ΔV) or over surfaces - as well as balance and moment equations of corresponding orders - are identical to each other. So, M. Beran [25] assumed that a spatial averaging is equivalent to a statistical averaging, isotropic over the ensemble, and therefore missed the specific features that are being considered here. However, we are interested in knowing if the assumption about their total equivalence can exclude significant physical effects or not.

In conventional turbulence theory we have [137] the postulate of commutability of averaging and differentiation operations

$$< \frac{\partial f}{\partial x} > = \frac{\partial}{\partial x} < f >, \tag{2.1.2}$$

Here $f = f(x, \chi)$ and the symbol <...> means any chosen averaging procedure, χ is the random parameter. Space averaging altogether with rescaling to macrocoordinates proves this claim and shows its sense.

Let us now consider the operation of spatial averaging (symbol $<f>$) of any plane flow along the axis x_j. It is easy to see that the left-hand side of expression (2.1.2) can be represented [220, 221] in the form

$$< \frac{\partial f}{\partial x_j} > = \frac{1}{\Delta X_j} \int_{X_j-(\Delta X_j/2)}^{X_j+(\Delta X_j/2)} \frac{\partial}{\partial x_j} < f >_j d\, x_j =$$

$$= \frac{< f >_j (X_j + \Delta X_j /2) - < f >_j (X_j - \Delta X_j /2)}{\Delta X_j} = \tag{2.1.3}$$

$$= \frac{\Delta < f >_j}{\Delta X_j}$$

Here $< f >_j =< f >_j (X_j)$ is the average value over the cross-section $\Delta S_j = \Delta X_k \Delta X_i$ of the selected axis j. The rule of commutability of averaging and differentiation operations follows from here, but this property is not the hypotheses but a consequence of change of description scale.

Actually, assume that $< f >_j$ does not already depend [220, 221] on random parameter (the averaging area ΔS_j includes the entire ensemble of realizations as the representative one) and

$$\Delta < f >_j / \Delta X_j \approx \partial < f >_j / \partial X_j$$

Then it is the differential operation *in a new scale*, and finally:

$$< \frac{\partial f}{\partial x_j} > = \frac{\partial < f >_j}{\partial X_j}. \qquad (2.1.4)$$

Thus, the spatial averaging procedure explains the postulate of commutability as a transition from the averaging over some volume $\Delta V = \Delta X_j \Delta S_j$ to that over the oriented area (symbol $< >_j$).

A similar problem appears in numerical modeling of equations in mathematical physics, when the finite-differential analogues of differential equations are developed. The resulting equations will correspond to new continuum equations in a larger scale as it was mentioned above. It is clear that the initial equations and the secondary ones can differ in the case of high spatial heterogeneity. The attempt to change turbulent parameters that were set by transition to another scale was made by A. Leonard [159] and discussed in the publications [208, 250].

Spatial averaging will lead also to volume average variables $< f >$ if we apply it to time derivatives assuming that this operation is done instantly, without changing time scale, that is,

$$< \frac{\partial f}{\partial t} > = \frac{\partial}{\partial t} < f > \qquad (2.1.5)$$

The divergent form of balance equations (1.2.1) - (1.2.3) (of mass, momentum, angular momentum and energy balances) led to values averaged over the volume ΔV and plane cross-sections $\Delta S_j = n_j \Delta S$ that composed the surface S, bounding this volume:

$$\frac{\partial}{\partial t}\int_{\Delta V} f\, dV = \int_S q_j^{(f)} n_j\, dS \tag{2.1.6}$$

Here $q_j^{(f)}$ is the corresponding flux. As it will be done below, the procedure of spatial averaging keeps the form of continuum equations because of (2.1.3) and (2.1.4). Then equation (2.1.5) will give

$$\frac{\partial < f >}{\partial t} = \frac{\partial < q_i^{(f)} >_i}{\partial X_i} \tag{2.1.7}$$

The volume average value $< f >$ and the flux, averaged at a surface $< q_j^{(f)} >_i$, appeared.

However, if we try to consider averaging over the volume ΔV of the velocity gradient (in other words, of the distortion rate tensor) then, according to (2.1.4), a new element of spatial averaging can be seen:

$$< \frac{\partial u_i}{\partial x_j} > = \frac{\partial U_i}{\partial X_j} + < \frac{\partial w_i}{\partial x_j} > = \frac{\partial U_i}{\partial X_j} + \frac{\partial < w_i >_j}{\partial X_j} \tag{2.1.8}$$

Actually, the stationary (spatially mean) part of local velocity u_i is singled out and we have to average the velocity pulsation w_i at plane cross-sections (in the Cartesian coordinates). Equality $U_i = <u_i>_i$ says nothing about

$< w_i >_j,\ i \neq j.$

If $< w_i >_j \neq 0$, then we have (no summing over j):

$$< \frac{\partial u_i}{\partial x_j} > = \frac{\partial U_i}{\partial X_j} + (1 - \delta_{ij})\frac{\partial}{\partial X_j} < w_i >_j, \tag{2.1.9}$$

On the other hand, let us consider the circulation Γ_i of the velocity u_j along the contour L_l $(l = j, k)$ that is in the form of a square, coinciding with a plane cross-section of the volume ΔV. Then

$$\Gamma_i = \int_{L_l} u_l \, dx_l = 2 \int\int_{\Delta S_j} \Phi_i \, dx_k \, dx_l = \int\int_{\Delta S_i} \varepsilon_{ijk} \frac{\partial u_k}{\partial x_j} dx_j \, dx_k = \quad (2.1.10)$$

$$= \varepsilon_{ijk} \left[\frac{\partial U_k}{\partial X_j} + \frac{\partial < w_k >_j}{\partial X_j} \right] \Delta S_i$$

One can see that at the right-hand side, the flux through ΔS_j of the mean vorticity Ω_i of the mean velocity field U_i, the surplus vorticity ω_i (its own total vorticity is $\Omega_i + \omega_i$) of velocity pulsation field w_k [11] appears:

$$\varepsilon_{ijk} \left[\frac{\partial U_k}{\partial X_j} + \frac{\partial < w_k >_j}{\partial X_j} \right] \Delta S_i = \Omega_i \Delta S_i + \sum_{k \neq i} < w_k >_j \Delta X_k \equiv$$

$$\Omega_i \Delta S_i + \sum_{k \neq i} v_k \Delta X_k = (\Omega_i + \omega_i) \Delta S_i , \quad (2.1.11)$$

Here we introduce a new vector v_k, - *averaged pulsation velocity along segment* ΔX_k whose circulation along the closed sum of such segments determines the spin vorticity flux[7].

We call ω_i the eddy *spin*.

So, in this situation there is a link between the circulation of a vector a_i along a contour C, and flux of pseudovector b_i through the surface S_C, stretched on this contour

$$\oint_C a_i \, d l_i = \int_{S_C} b_i \, n_i \, d S \quad (2.1.12)$$

According to the Stokes theorem, the line averaging of a vector along a closed contour, made of edges of the elementary volume, results in the total flux of vortexes through the corresponding cross-section.

Let us consider[8] the oriented contour C, composed of the segments (δ_{ij} is the unit tensor):

---

[7] Another example could be the Lagrange averaging for a tracer motion, see [163].

$$[X_i, X_i + \delta_{i1} \Delta X_1],$$

$$[X_i + \delta_{i1} \Delta X_1, X_i + \delta_{i1} \Delta X_1 + \delta_{i2} \Delta X_2],$$

$$[X_i + \delta_{i1} \Delta X_1 + \delta_{i2} \Delta X_2, X_i + \delta_{i2} \Delta X_2],$$

$$[X_i + \delta_{i2} \Delta X_2, X_i].$$

Let us consider the circulation of a velocity vector $\boldsymbol{u}$ along this contour

$$\oint_C u_i \, dl_i = \langle u_1 \rangle_1 \Delta X_1 + \left(\langle u_{22} + \frac{\partial \langle u_2 \rangle_1}{\partial X_1} \Delta X_1 \right) \Delta X_2 -$$

$$-\left(\langle u_1 \rangle_1 + \frac{\partial \langle u_1 \rangle_2}{\partial X_2} \Delta X_2 \right) \Delta X_1 - \langle u_2 \rangle_2 \Delta X_2 =$$

(2.1.13)

$$\left(\frac{\partial \langle u_2 \rangle_1}{\partial X_1} - \frac{\partial \langle u_1 \rangle_2}{\partial X_2} \right) \Delta X_1 \Delta X_2 =$$

$$\varepsilon_{3jk} \frac{\partial \langle u_k \rangle_j}{\partial X_j} \Delta X_1 \Delta X_2$$

Introducing here the formulae for local velocity

$$\langle u_i \rangle_{\mathrm{k}} = \mathrm{U_i} + \langle w_i \rangle_k, \qquad i \neq k. \tag{2.1.14}$$

we obtain

---

[8] O. Dinariev. Private communication (2002)

$$\oint_C w_i \, dl_i \equiv \oint v_i \, dl_i = \varepsilon_{3jk} \frac{\partial v_k}{\partial X_j} \Delta X_1 \Delta X_2 \qquad (2.1.15)$$

Therefore the circulation of pulsation velocity determines the flux of surplus vortexes, consequently. The averaged distortion rate tensor includes the average velocity rotation and the circulation of the pulsation velocity along the contour of area ΔS_j.

Let us consider now the volume averaging of the viscous stress tensor (1.1.11). We see that the symmetrical part of the distortion rate does enter the average viscous stresses, averaged also over the elementary volume:

$$< \sigma_{ij} > = -p + \frac{2}{3} \rho \nu < \frac{\partial u_k}{\partial x_k} > \delta_{ij} +$$
$$\rho \nu (\frac{\partial U_i}{\partial X_j} + \frac{\partial U_j}{\partial X_i}) + \rho \nu (\frac{\partial < w_i >_j}{\partial X_j} + \frac{\partial < w_j >_i}{\partial X_i}), \qquad (2.1.16)$$

We get that the mean strain rate is determined by averaged velocities and by the following additional term:

$$\rho \nu (\frac{\partial < w_i >_j}{\partial X_j} + \frac{\partial < w_j >_j}{\partial X_i}) \equiv$$
$$\rho \nu (\frac{\partial v_i}{\partial X_j} + \frac{\partial v_j}{\partial X_i}) = < \sigma_{ij} > * \qquad (2.1.17)$$

The latter reflects the influence of turbulent eddies on viscous dissipation in a flow (compare with the known Einstein formula [73] that accounts for effect of the concentration of diluted solid suspension, see Section 4.2).

2.2. CHAIN OF TURBULENCE BALANCE EQUATIONS

Let us reproduce here integral balances (1.1.27) - (1.1.29) for the elementary macrovolume ΔV. They have the following form:

$$\frac{\partial}{\partial t} \int_{\Delta V} \rho \, dV + \int_{\Delta S} \rho \, u_i \, dS_j = 0, \qquad (2.2.1)$$

$$\frac{\partial}{\partial t}\int_{\Delta V}\rho\, u_i\, dV + \int_{\Delta S}\rho\, u_i\, u_j\, dS_j = \int_{\Delta S}\sigma_{ij}\, dS_j + \int_{\Delta V} F_i\, dV \tag{2.2.2}$$

$$\frac{\partial}{\partial t}\int_{\Delta V}\rho(\mathrm{E}+\frac{u_i\, u_i}{2})dV + \int_{\Delta S}\rho(\mathrm{E}+\frac{u_i\, u_i}{2})\, u_j\, dS_j = \tag{2.2.3}$$

$$= \int_{\Delta S}\sigma_{ij}\, u_i\, dS - \int_{\Delta S} q_j\, dS_j + \int_{\Delta V} Q dV + \int_{\Delta V} F_i\, u_i\, dV$$

Here the surface ΔS bounds the volume ΔV.

Due to continuity of all variable fields in a viscous fluid, we can apply the Ostrogradskiy-Gauss theorem. Then [212-214], dividing the integral balances by the value ΔV, in accordance with Section 2.1, we represent the latter balances in a form of macrodifferential equations:

$$\frac{\partial <\rho>}{\partial t} + \frac{\partial <\rho\, u_j>_j}{\partial X_j} = 0, \tag{2.2.4}$$

$$\frac{\partial <\rho\, u_i>}{\partial t} + \frac{\partial <\rho\, u_i\, u_j>_j}{\partial X_j} = \frac{\partial <\sigma_{ij}>_j}{\partial X_j} + <F_i>, \tag{2.2.5}$$

$$\frac{\partial}{\partial t}<\rho(\mathrm{E}+\frac{u_i u_i}{2})> + \frac{\partial}{\partial X_j}<\rho(\mathrm{E}+\frac{u_i u_i}{2})u_j>_j =$$

$$= \frac{\partial}{\partial X_j}<\sigma_{ij} u_i>_j + <F_i u_i> + \frac{\partial <q_j>_j}{\partial X_j} + <Q>, \tag{2.2.6}$$

In (2.2.4) the volume and surface averaged values have appeared simultaneously:

$$< \rho >= \frac{1}{\Delta V} \int_{\Delta V} \rho \, dV,$$

$$< \rho u_j >_j = \frac{\Delta X_j}{\Delta V} \int_{\Delta S_j} \rho u_j \, dS_j , \tag{2.2.7}$$

$$\Delta S_j = \Delta X_k \Delta X_l , \qquad j \neq k \neq l .$$

The elementary volume ΔV itself, as well as the area of these limiting cross-sections, are assumed to be big enough to include the ensemble of all random realizations of turbulent mesovortexes.

Let condition $\Delta >> \lambda$ be valid in the volume ΔV. This means that the averaged values

$$< \rho >, \qquad < \rho u_j >_j$$

as well as other mean variables are not anymore random but determined and regular functions of only macrocoordinates and they are attached to the mass center X_i of the elementary volume.

The first term of (2.2.5) corresponds to the usual representation of the average velocity U_i through local impulse in the volume ΔV :

$$< \rho > U_i = < \rho u_i > \tag{2.2.8}$$

According to (2.2.7) the average velocity is determined by the mass flow through the cross-section at any macro point:

$$U_i = \frac{< \rho u_i >_i}{< \rho >} \tag{2.2.9}$$

Usually it is assumed that (2.2.8) and (2.2.9) yield the same value.

The velocity turbulent field is represented thereby in the form of a sum of regular and pulsation terms:

$$u_i (x_j , t) = U_i (x_j , t) + w_i(x_j , t; \chi) \tag{2.2.10}$$

The pulsation velocity w_i is a function of a space point and the "random parameter" χ. In accordance with (2.2,9) we have:

$$< \rho w_i >_i = 0. \tag{2.2.11}$$

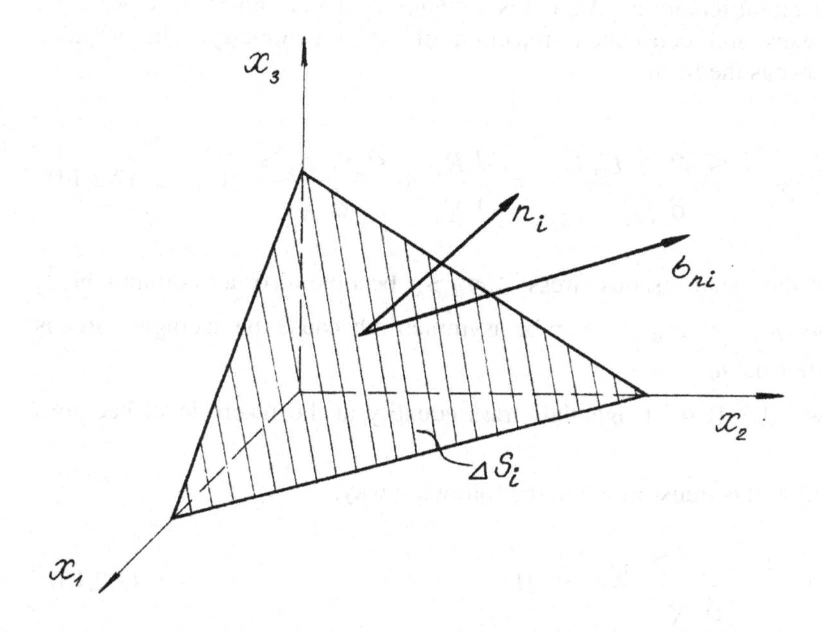

2.1.1. The Cauchy tetrahedron illustrates stress σ_{ni} in an oriented plane

However, tangential averaging over the area $\Delta S_j = \Delta X_k \Delta X_i, \quad j \neq k \neq i$ (see Sections 2.1 and 5.4) will determine the vector that had appeared in the distortion rate tensors:

$$< \rho w_i >_j = v_i.$$

Considering now the momentum flux:

$$< \rho\, u_i\, u_k >_k =< \rho > U_i U_k + < \rho\, w_i w_k >_k \tag{2.2.12}$$

we see the Reynolds stresses:

$$R_{ik} = - < \rho\, w_i w_k >_k \,. \tag{2.2.13}$$

The Reynolds stresses are determined by transfer of impulse pulsation $\rho\, w_i$ with velocity w_i along the normal n_k of the oriented cross-section. It exactly corresponds to the Cauchy initial concept.[9] of continuum mechanics.

The most essential feature of (2.2.13) is violation of the commutability principle for tensor indexes and consequent rejection of stress symmetry. The impulse balance equation has the form

$$\frac{\partial < \rho > U_i}{\partial t} + \frac{\partial < \rho > U_i U_j}{\partial X_j} = \frac{\partial R_{ij}}{\partial X_j} + \frac{\partial < \sigma_{ij} >_j}{\partial X_j} + F_i \,, \tag{2.2.14}$$

The indexes of the mean viscous stress $< \sigma_{ik} >_k$ become also not commutable[10]. So, both tensors R_{ik} , $< \sigma_{ik} >_k$ can be asymmetric because the averaging area is oriented by its normal n_k .

Therefore "law" (1.1.4) for tangential stress equality at the macro-level becomes invalid.

Mass balance (2.2.1) is transformed in the following way:

$$\frac{\partial < \rho >}{\partial t} + \frac{\partial < \rho > U_j}{\partial X_j} = 0. \tag{2.2.15}$$

Expressions (2.2.8) and (2.2.9), which were accounted for by (2.2.14), (2.2.15), state the equality of two different velocity definitions - by the impulse in the elementary volume and by mass flux. This assumes the equivalency of volume and cross-section averaging of the vector field function:

$$U_i = < u_i > \equiv < u_i >_i \qquad \rho = const \tag{2.2.16}$$

However, the volume averaged tensors of the second rank:

---

[9] Here terms of the mean velocity field that have the order $O(\lambda^2/\Delta X^2)$ are omitted. Leonard suggested [159] to hold such terms to account for eddies larger than λ, see (1.2.3).

[10] Ahmadi et al [4] mentioned that mean *viscous* stress tensor depended on the orientation of a cross-section. A. Zagustin [302] had noted the same for *Reynolds* stresses.

$$< \rho u_i u_j > , < \sigma_{ij} >$$

are just parts of true stress tensors introduced as forces acting at oriented cross-sections of the turbulent medium and appearing in impulse balance (2.2.14).

Actually, by volume integrating of (2.2.2), we obtain:

$$< \rho u_i u_k - \sigma_{ik} >= \frac{\partial}{\partial t} < \rho u_i x_k > + \frac{\partial}{\partial X_j} < \rho u_i u_j x_k >_j - \quad (2.2.17)$$

$$- \frac{\partial}{\partial X_j} < \sigma_{ij} x_k >_j - < F_i x_k > .$$

Multiply now equation (2.2.14) by X_k (this coordinate of the mass center can be introduced under the averaging sign as the constant value in the elementary volume):

$$< \rho u_i u_k - \sigma_{ik} >_k = \frac{\partial}{\partial t} < \rho u_i X_k > +$$

$$\frac{\partial}{\partial X_j} < \rho u_i u_j X_k > - \frac{\partial}{\partial X_j} < \sigma_{ij} X_k >_j - F_i X_k \quad (2.2.18)$$

The difference of these two expressions yields:

$$N_{ik} =< \sigma_{ik} > - < \rho u_i u_k > + \frac{\partial}{\partial t} < \rho u_i \xi_k > + \quad (2.2.19)$$

$$\frac{\partial N_{ijk}}{\partial X_j} < \rho u_i u_j \xi_k - \sigma_{ij} \xi_k >_j - < F_i \xi_k >$$

Here $\xi_j = x_j - X_j$ is a radius-vector relative to the mass center of the elementary volume and the impulse flux is given by

$$N_{ik} =< \sigma_{ik} - \rho u_j u_k >_k \quad (2.2.20)$$

As follows from the right-hand side of equation (2.2.19), the value of the total momentum flux N_{ij} includes the stress tensors, averaged over the volume ΔV, but as a part of true macro-stress tensor. In accordance with the Cauchy concept (see [217, 287]), the latter corresponds exactly to a force, acting at the oriented area. Therefore, a volume-average stress tensor cannot be used as a macro-stress [217] instead of N_{ij} in the impulse balance (compare with [22]).

One may see that additional terms of the right-hand side of (2.2.19) include the moments of velocity fields of higher order. They can be expressed in their turn by the balance equations for the moments of momentum of the next orders, etc. Indeed, the impulse flux N_{ik} is expressed by (2.2.19) that includes a space derivative of the third order stress moment:

$$N_{ijk} = <(\sigma_{ij} - \rho u_i u_j)\xi_k>_j$$

The corresponding equations are obtained as result of multiplying of (2.2.2) by $x_k x_m \cdots$, followed by the volume averaging. The subtraction of their analog, obtained under multiplying by the $X_k X_m \cdots$ values, yields the chain of equations [161, 217] of the moments of increasing order

$$N_{ijkm\cdots} = <(\sigma_{ij} - \rho u_i u_j)\xi_k \xi_m \cdots>_j \tag{2.2.21}$$

They are averaged over all the oriented cross-sections:

$$<(\sigma_{ik} - \rho u_i u_k)\xi_m \cdots>_k = <(\sigma_{ik} - \rho u_i u_k > \xi_m \cdots >$$

$$+\frac{\partial}{\partial t} < \rho u_i \xi_k \xi_m \cdots > - \frac{\partial}{\partial X_j} < \sigma_{ij} \xi_k \xi_m \cdots - \rho u_i u_j \xi_k \xi_m \cdots >_j - \tag{2.2.22}$$

$$< F_i \xi_k \xi_m \cdots > .$$

Everywhere, the second index corresponds to the normal of the mentioned cross-section. Besides, the Mindlin double stresses [187] as gradient of stresses may appear in this chain.

In general, the *n*-order moment is the dyadic product of the *(n - 1)*-order moment ($n \geq 1$) and of a radius-vector of the point under consideration (relative to the mass center of elementary macrovolume, corresponding to the coordinate system). Such a consequence of equations is an analog of the known Friedman-Keller's chain [91] where the Navier-Stokes equation was multiplied by local velocity vectors u_i, $u_i u_k$,... and averaged by statistical isotropic procedure.

Chain (2.3.5), interrupted at the nth step, is needed in the introduction of the closing relations (and it is usual for the theory of turbulence). In the statistical theory of turbulence the closing relations are found in the form of interconnections between the correlation moments of lower and higher ranks. In the case of spatial averaging, suggested here, the completion is carried out in accordance with the constitutive (in other words, "half-empirical") rules of continuum mechanics following the Boussinesq approach [68, 96]. Now this choice has to be made consistent with the thermodynamics of irreversible processes. This procedure will be illustrated below.

Moreover, referring to numerical calculation methods, where spatial averaging was used in practice, we should note that those algorithm schemes would be the best, where a higher number of equations from balance chain (2.3.5) is used. As to the difference in the rules of averaging for different values, it has appeared already in the theory of turbulence (remember the different Lagrange and Euler scales of correlation [163]).

Particularly, in the most important case, the chain breaking happens on the third order equation, i.e. at the angular momentum balance.

According to turbulence theory, a commutative principle for indexes in tensors of one-point velocity correlations [20] and the identification of the latter with the Reynolds stress tensor would interrupt the moment chain. The decisive counter argument consists in the averaging over an oriented cross-section as the only one proper for stresses. This rule is automatically implemented under the application of spatial averaging to the initial Navier-Stokes equations.

However, in turbulence the corresponding internal mesostructure is generated by boundaries and, therefore, these "half-empirical" completing laws have to be determined really empirically. Correspondingly, only general features of the constitutive laws can be established a priori on the base of reasonable kinetic assumptions.

2.3. ANGULAR MOMENTUM BALANCE

Equation (2.2.19) multiplied by the Levi-Civita alternating tensor ε_{ijk} yields the angular momentum balance about a mass center X_i of the volume ΔV in a form:

$$\frac{\partial}{\partial t} < \varepsilon_{ijk}\, \rho\, u_k\, \xi_j > + \frac{\partial}{\partial X_j} < \varepsilon_{ilk}\, \rho\, u_k\, \xi_l\, u_j >_j +$$

$$\varepsilon_{ilk} < \rho\, u_k\, u_l >_l = \frac{\partial}{\partial X_j} < \varepsilon_{ilk}\, \sigma_{kj}\, \xi_l >_j + \tag{2.3.1}$$

$$\varepsilon_{ilk} < \sigma_{kl} >_l + < \varepsilon_{ilk}\, \xi_l\, F_k > .$$

On the other hand, the local velocity field is represented as:

$$u_i(x_j, t) = U_i(x_j, t) + w_i(x_j, t\,;\chi) = U_i(X_j, t) +$$

$$(\partial U_i \,/\, \partial X_j)\xi_j + O\,(\Lambda^2 / \,L^2) + w_i(X_j + \xi_j, t\,;\chi). \tag{2.3.2}$$

Symbol χ underlines randomness of pulsations.

We assume that volume ΔV is filled totally with n eddies, the velocity pulsation field inside λ - vicinity of ξ_m - center of each eddy being represented as:

$$w_k = \overline{w}_k + \frac{\partial \overline{w}_k}{\partial \overline{\xi}_m} \zeta_m, \quad \zeta_m = \xi_m - \overline{\xi}_m, \tag{2.3.3}$$

Here $\overline{w}_k$ is the eddy mass center velocity and

$$0 \leq |\zeta_m| \leq \lambda$$

Formula (2.3.3) means the eddy modeling by a solid-body distortion rate, that is, the Prandtl's "mole" ideology [96, 239] is used.

The left-hand side of (2.3.1) contains the rate of local change of *angular momentum* of a fluid element:

$$M_i = \varepsilon_{ilk} \; \rho \; u_k \; \xi_l$$

but averaged over an elementary macrovolume. For velocity field (2.3.2) the spatial averaged angular momentum is defined and transformed, consequently, as follows:

$$< M_i >=< \varepsilon_{ilk} \, \rho \, u_k \, \xi_l >=$$

$$\varepsilon_{ilk} \frac{\partial U_k}{\partial X_m} < \rho \, \xi_m \, \xi_l > + < \varepsilon_{ilk} \, \rho \, w_k \, \xi_l >= \qquad (2.3.4)$$

$$\varepsilon_{ilk} \frac{\partial U_k}{\partial X_m} I_{ml} + < \varepsilon_{ilk} \, \rho \, \overline{w_k} \, \overline{\xi_l} > + < \varepsilon_{ilk} \, \Phi_{km} \, i_{ml} > .$$

Here I_{ml} ($\sim \Delta X^2$) is a scaled inertia moment of fluid filling the elementary volume ΔV, d_{km} is the local distortion rate, ρi_{ml} is the inertia tensor, corresponding to the subscale eddy volume, ρJ_{ml} is its mean value:

$$\rho I_{ml} =< \rho \, \xi_m \, \xi_l >= \frac{1}{\Delta V} \int_{\Delta V} \rho \, \xi_m \, \xi_l \, \Delta V,$$

$$2\Phi_{km} = \frac{\partial U_k}{\partial X_m} + \frac{\partial \overline{w_k}}{\partial \overline{\xi}_m}, \qquad J_{km} =< i_{km} > \qquad (2.3.5)$$

$$\rho \, i_{ml} = \frac{1}{\Delta V_\lambda} \int_{\Delta V_\lambda} \rho \, \zeta_m \, \zeta_l \, dV, \quad J'_{km} = i_{km} - J_{km}$$

Thus, the first term of $< M_i >$, according to the right-hand part of (2.3.4), is the angular momentum of the average velocity field in the volume ΔV.

The second one is the mean angular momentum of some irregular eddies of λ - scale that fill a whole ΔV. This term can be taken into account for the analysis of turbulent flows, stationary in mean, when the simultaneous averaging over the large time interval is carried out. However, we shall neglect it in the present context.

The third one is the moment corresponding to small-scale energy contained turbulence.

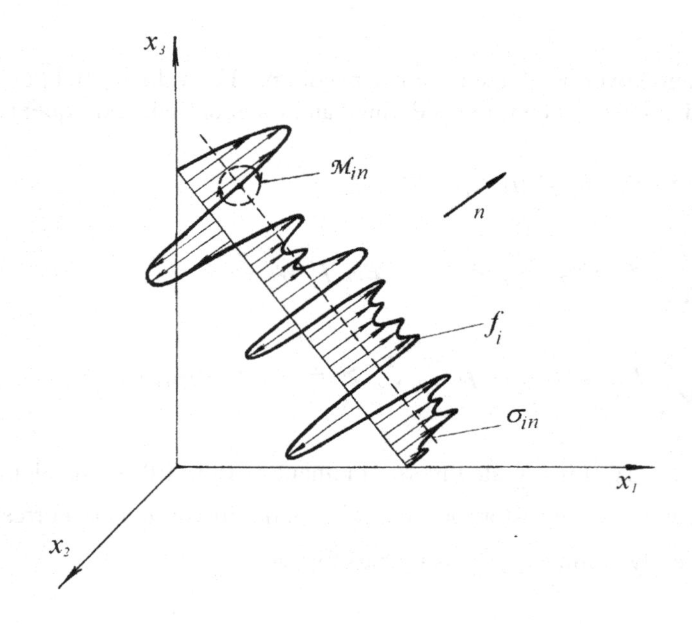

Figure 2.3.1. Eddies generate stress irregularity and couple stresses at a cross-section of turbulent flow

Representation (2.3.4) for the average angular momentum can be written now in the form:

$$< M_i >= \varepsilon_{ilk}\, \rho\, I_{ml} \frac{\partial U_k}{\partial X_m} + \varepsilon_{ilk}\, \rho J_{ml} (\frac{\partial U_k}{\partial X_m} + \omega_{km}) \qquad (2.3.6)$$

$$\omega_{km} = 2< \Phi_{km} > - \partial U_k / \partial X_m ,$$

its pulsation $M_i^{/}$ being determined by the formula:

$$M_i^{\prime} = M_i - < M_i >=$$

$$2\, \varepsilon_{ilk}\, \Phi_{km}\, \rho (J_{ml} + J_{ml}^{\prime}) - 2\, \varepsilon_{ilk}\, \rho J_{ml} < \Phi_{km} >= \tag{2.3.7}$$

$$2 \varepsilon_{ilk}\, \rho\, J_{ml}\, \Phi_{km}^{\prime} + 2 \varepsilon_{ilk}\, \rho\, J_{ml}^{\prime} < \Phi_{km} >,$$

Let us introduce the *viscous couple stresses* $m_{ij} =< \varepsilon_{ilk}\, \sigma_{kj}\, \xi_l >_j$ that are connected with an irregularity of viscous microstress distribution at a cross-section of the elementary volume ΔV.

The *turbulent couple stresses*, caused by the pulsation transfer of the eddy's angular momentum fluctuations, are introduced as:

$$\mu_{ij} =< - M_i^{\prime}\, w_j >_j = -2\, \varepsilon_{ilk}\, \rho\, J_{ml} < \Phi_{km}^{\prime}\, w_j >_j$$

$$- 2 \varepsilon_{ilk} < \Phi_{km} >< \rho\, J_{ml}^{\prime}\, w_j >_j , \tag{2.3.8}$$

The second term of (2.3.8) will be omitted. Then angular momentum balance (2.3.1) is reduced to the Cosserat equation [54, 125]

$$\frac{\partial < M_i >}{\partial t} + \frac{\partial < M_i >_j U_j}{\partial X_j} = \tag{2.3.9}$$

$$\frac{\partial \mu_{ij}}{\partial X_j} + \frac{\partial m_{ij}}{\partial X_j} + \varepsilon_{ilk} (R_{lk} + < \sigma_{kl} >_l) + C_i$$

Now our aim is to exclude the angular momentum, corresponding to total fluid mass in ΔV, from balance (2.3.9).

We shall assume that this angular momentum is mainly determined by vorticity of the mean velocity field, i.e. here we shall neglect the contribution of deformation rate:

$$\frac{\partial U_i}{\partial X_k} = \frac{1}{2} \varepsilon_{kim}\, \Omega_m + \frac{1}{2} (\frac{\partial U_i}{\partial X_j} + \frac{\partial U_j}{\partial X_i}) \approx \frac{1}{2} \varepsilon_{kim}\, \Omega_m . \tag{2.3.10}$$

Then the corresponding part $< M_i >^{\Omega}$ of angular momentum (2.3.4) is transformed in the following way:

$$< M_i >^{\Omega} = \varepsilon_{ilk} I_{ml} \frac{\partial U_k}{\partial X_m} \approx \frac{1}{2} \varepsilon_{ilk} \varepsilon_{kmp} I_{ml} \Omega_p \tag{2.3.11}$$

If one applies the vorticity operation to the average velocity field, we obtain the known equation of the average eddy diffusion

$$\frac{\partial \Omega_p}{\partial t} + \frac{\partial \Omega_p U_j}{\partial X_j} =$$

$$\Omega_j \frac{\partial U_p}{\partial X_j} + \frac{1}{2\rho} \varepsilon_{pki} \frac{\partial F_i}{\partial X_k} + \frac{1}{2\rho} \frac{\partial^2}{\partial X_k \partial X_j} \varepsilon_{pki} R_{ij} \tag{2.3.12}$$

Let us now use evolution equation (1.3.22) of the inertia moment I_{km}, written for symmetrical volume ΔV:

$$\varepsilon_{ijk} \frac{\partial U_k}{\partial X_j} \left(\frac{\partial I}{\partial t} + \frac{\partial I U_j}{\partial X_j} \right) = -I \Omega_j \frac{\partial U_i}{\partial X_j};$$

$$I_{ml} = I \delta_{ml} \tag{2.3.13}$$

Equation (2.3.12), multiplied by the value $I \delta_{ml} \varepsilon_{ilk} \varepsilon_{kmj} \equiv 2 I \delta_{km}$ after the following summation with expression (2.3.13), yields:

$$\frac{\partial}{\partial t} < M_i >^{\Omega} + \frac{\partial}{\partial X_j} < M_i >^{\Omega}_j U_j = \tag{2.3.14}$$

$$\frac{I}{\rho} \varepsilon_{ikp} \left(\frac{\partial F_p}{\partial X_k} + \frac{\partial^2 R_{ij}}{\partial X_k \partial X_j} \right)$$

If one assumes (compare with [212]) that

$$C_i^\omega = C_i - \frac{I}{\rho} \varepsilon_{ikp} \left(\frac{\partial F_p}{\partial X_k} + \frac{\partial^2 R_{ij}}{\partial X_k \partial X_j} \right) \tag{2.3.15}$$

the difference between (2.3.8) and (2.3.14) allows us to find the expression for kinetic moment $<M_i>^\omega$, determined just by the rotation rate of turbulent moles:

$$\frac{\partial}{\partial t} < M_i >^\omega + \frac{\partial}{\partial X_j} < M_i >^\omega{}_j U_j =$$
$$\frac{\partial \mu_{ij}}{\partial X_j} + \frac{\partial m_{ij}}{\partial X_j} + \varepsilon_{ilk} (R_{lk} + < \sigma_{kl} >_l) + C_i^\omega \tag{2.3.16}$$

Here C_i^ω is the bulk external force moment and (2.3.16) coincides with angular momentum balance, which is conventional for polar or, otherwise, asymmetrical hydrodynamics:

It will be useful for further consideration to exploit simplifying suppositions in expression (2.3.9) for the turbulent couple stresses:

$$\mu_{ij} =< - M_i' w_j >_j = - \varepsilon_{ilk} \varepsilon_{kmp} \rho J_{ml} < 2 \Phi'_p w_j >_j -$$
$$\frac{1}{2} \varepsilon_{ilk} \varepsilon_{kmp} (\Omega_p + \omega_p) < \rho J'_{ml} w_j >_j \tag{2.3.17}$$

We have used again the tensor relation (1.3.21):

$$\varepsilon_{ijk} \varepsilon_{lpk} = \delta_{il} \delta_{jp} - \delta_{ip} \delta_{jl} .$$

2.4. EVOLUTION OF MOMENT OF INERTIA

The moment of inertia evolution equation can be found as follows. Multiplying the mass balance equation (2.1.1) by $\xi_k \xi_m$ in the dyad manner, we obtain:

$$\frac{\partial \rho\, \xi_k \xi_m}{\partial t} + \frac{\partial}{\partial x_j} \rho\, u_j \xi_k \xi_m = \rho\, u_j \xi_m \delta_{kj} + \rho\, u_j \xi_k \delta_{mj} \qquad (2.4.1)$$

Averaging the result over the ΔV gives us [215]:

$$\frac{\partial}{\partial t} < \rho \xi_k \xi_m > + \frac{\partial}{\partial X_j} < \rho \xi_k \xi_m u_j > =$$
$$< \rho u_k \xi_m > + < \rho u_m \xi_k > . \qquad (2.4.2)$$

To find the inertia moments connected with the eddy motion inside the volume ΔV, considered above, we represent again the radius-vector ξ_m in the form of

$$\xi_m = \overline{\xi_m} + \zeta_m .$$

Then the inertia moment of the volume $\Delta V = n \Delta V_\lambda$, where ΔV_λ is the subscale eddy volume, is transformed in the following manner:

$$\rho I_{km} = < \rho\, \xi_k \xi_m > = \frac{1}{\Delta V} \sum_{\lambda=1}^{n} \rho\, \xi_k \xi_m \Delta V_\lambda + \frac{\Delta V_\lambda}{\Delta V} \sum_{l=1}^{n} \frac{1}{\Delta V_\lambda} \times$$
$$\int_{\Delta V_\lambda} \rho\, \zeta_k \zeta_m dV = \frac{1}{\Delta V} \sum_{\lambda=1}^{n} (\rho\, \overline{\xi}_k \overline{\xi}_m \Delta V_\lambda) + \rho\, J_{km}; \qquad (2.4.3)$$

$$J_{km} = \frac{1}{n} \sum_{\lambda=1}^{n} i_{km}$$

The mean (for the volume ΔV) value of the mole inertia moment has the order of:

$$J_{km} \sim \lambda^2 \sim (\Delta X)^2 / n.$$

If $n \to \infty$, the mesostructure disappears and $J_{km} \to 0$.

However, such a limit transition is not permissible [222] if one intends to account for turbulent meso-vortexes. That is why we use the equality

$$I_{km} = I^{\omega}_{km} + J_{km}, \qquad (2.4.4)$$

Here I^{ω}_{km} is a scaled moment of inertia for the small-scale eddies mass-centers in the elementary volume, and $I^{\omega}_{km} \to I_{km}$ when $n \to \infty$.

Besides we can use representations (2.3.2), (2.3.3) and (2.3.5). Equation (2.4.2) is transformed, correspondingly, into the next one:

$$\frac{\partial}{\partial t}(I^{\omega}_{km} + J_{km}) + \frac{\partial}{\partial X_j}(I^{\omega}_{km} + J_{km})U_j =$$

$$I^{\omega}_{nm}\frac{\partial U_k}{\partial X_n} + I^{\omega}_{kn}\frac{\partial U_m}{\partial X_n} + 2 < \Phi_{kn}\, i_{mn} + \Phi_{mn}\, i_{nk} > - \qquad (2.4.5)$$

$$\frac{\partial}{\partial X_j} < (I_{km} + i_{km})\,\omega_j >_j , \qquad \omega_i = \varepsilon_{ijk}\,\omega_{jk}$$

From another point of view, one can similarly (or by using[11] the method [210]) obtain the balance correlation for the inertia moment I_{km} of the fluid inside the elementary volume:

$$\frac{\partial I_{km}}{\partial t} + \frac{\partial I_{km} U_j}{\partial X_j} = \frac{\partial U_k}{\partial X_n} I_{nm} + \frac{\partial U_m}{\partial X_n} I_{nk} . \qquad (2.4.6)$$

With the assumption that $I_{km} \approx I^{\omega}_{km}$, the difference between balances (2.4.5) and (2.4.6) yields the evolution of the turbulent field inertia moment

$$\frac{\partial}{\partial t} J_{km} + \frac{\partial}{\partial X_j} J_{km} U_j = 2 < \Phi_{kn}\, i_{mn} + \Phi_{mn}\, i_{nk} > - \qquad (2.4.7)$$

$$\frac{\partial}{\partial X_j} < i_{km}\,\omega_j >_j ,$$

[11] The inaccuracy in the inertia moment definition is compensated by the corresponding changes of eddy's angular velocity to keep the moment M_i value.

Here the condition $< I_{km} w_i >_i = I_{km} < w_i >_i = 0$ was used. Then introducing the pulsation J'_{km} of mole moment of inertia and pulsation Φ'_{km} of the angular velocity tensor:

$$i_{km} = J'_{km} + J_{km}, \quad \Phi_{kn} = \Phi'_{kn} + \frac{1}{2}\left(\frac{\partial U_k}{\partial X_n} + < \frac{\partial \overline{w_k}}{\partial \xi_n} >\right),$$

we transform evolution equation (2.4.7) in the following manner:

$$\frac{\partial J_{km}}{\partial t} + \frac{\partial J_{km} U_j}{\partial X_j} = 2 J_{kn} \Phi_{mn} + 2 J_{mn} \Phi_{kn} + Q_{km} + \frac{\partial \Pi_{kmj}}{\partial X_j} \qquad (2.4.8)$$

The pulsation source (sink) and diffusion spreading of the moment of inertia of a turbulent eddy are taken here into account for:

$$Q_{km} =< 2\, \Phi'_{kn} J'_{nm} >, \qquad \Pi_{kmj} = - < J'_{km} w_j >_j . \qquad (2.4.9)$$

However, for simplicity we often reject the right-hand part of (2.4.8).

The hypothesis is that the elliptical Prandtl mole can model turbulent eddy and this seems to be a quite sufficient assumption about its geometry. Vector n_m, directed along the bigger ellipsoid's axis, can characterize such geometry.

Accordingly, the inertia moment of the turbulent eddy proper has a form

$$\rho J_{ml} = \rho J_{\perp} \delta_{ml} + \rho (J_{\_} - J_{\perp}) \xi_m \xi_l .$$
(2.4.10)

The other vector values, characterizing the turbulent macropoint, are the translation U_i and angular $\Omega_i / 2$, $\phi_i = \omega_i / 2$ velocities. The angular velocities are the pseudo-vectors and they characterize the "eddy" anisotropy of the flow.

Traditional statistical analysis allows us to study the correlations of a velocity pulsations field (i.e. actually the most probable eddy structure which has the λ scale of the turbulent mole). It will be done in Chapter 6.

Consider the case where vectors of the total angular velocity $(\Omega_i + \omega_i) / 2$ and mean vector Ω_i (as well as the vector of the mole's anisotropy n_i) have one non-zero component, being orthogonal to the stream plane. Such plane flows permit a simple account for the spin rotation of eddies (moles):

$$\vec{U} = \vec{U}\ (U_1, U_2, 0), \qquad \vec{\xi} = \vec{\xi}\ (0,\ 0,\ 1),$$
$$\vec{\omega} = \vec{\omega}(\ 0,\ 0,\ \omega), \qquad \vec{\Omega} = \vec{\Omega}\ (0,\ 0,\ \Omega). \tag{2.4.11}$$

This yields the following simplifications for the mole's *geometrical moment* that appeared in the inertia moment:

$$J_{ml} = J_{\perp}\,\delta_{ml}\,,\ m,\ l = 1,\ 2; \qquad J_{ml} = 0,\ m \neq l,$$
$$J_{ml} = J\,\delta_{ml}\,, \qquad m = l = 3, \qquad J = J_{-}\,. \tag{2.4.12}$$

Equation (2.3.17) of the mole's angular momentum becomes simpler if $J_{km} = J\ \delta_{km}$:

$$\rho\frac{\partial}{\partial t}J(\Omega + \omega) + \rho\frac{\partial}{\partial X_j}J(\Omega + \omega)U_j =$$
$$\frac{\partial \mu_{ij}}{\partial X_j} + \varepsilon_{ilk}\,R_{lk} \tag{2.4.13}$$

Here $j = 1,\ 2$ and the effects of the outer volume moment C_i^{ω} are rejected as well as the couple *viscous* stresses m_{ij} .

Consider the turbulent flow to be incompressible:

$$\frac{\partial U_1}{\partial X_1} + \frac{\partial U_2}{\partial X_2} = 0. \tag{2.4.14}$$

Impulse balance (2.2.14) also becomes simpler:

$$\frac{\partial U_i}{\partial t} + (U_1\frac{\partial}{\partial X_1} + U_2\frac{\partial}{\partial X_2})U_i = \tag{2.4.15}$$

$$-\frac{1}{\rho}\frac{\partial p}{\partial X_i} + \frac{1}{\rho}\left(\frac{\partial R_{i1}}{\partial X_1} + \frac{\partial R_{i2}}{\partial X_2}\right)$$

Let us write geometrical moment evolution (2.4.8) in the following form

$$\frac{\partial J}{\partial t} + \frac{\partial J U_j}{\partial X_j} = \frac{\partial \Pi_j}{\partial X_j},$$

$$\Pi_j = - < J' w_j >_j \tag{2.4.16}$$

Here Π_j is the diffusion flux of the moment of inertia [213, 214, 222], caused by the correlation of the pulsation velocities and the inertia moment deviations J' from its mean value J.

The inertia moment has the order of internal length square $J \sim \lambda^2$. Therefore equation (2.4.16) may be interpreted as the evolution of the internal length scale λ, that is, of turbulent mesostructure:

$$\frac{\partial \lambda^2}{\partial t} + \frac{\partial \lambda^2 U_j}{\partial X_j} = \frac{\partial \Pi_j}{\partial X_j}, \qquad \Pi_j = - < (\lambda^2)' w_j >_j \tag{2.4.17}$$

CHAPTER 3

TURBULENCE AS OPEN THERMODYNAMIC SYSTEM

3.1. ENERGY AND ENTROPY BALANCES

Let us consider the total energy balance, represented by the following set of equations: (1.1.5) - for the microlevel, (1.1.29) - for arbitrary volume V, (2.2.6) - for differential volume ΔV (in macrocoordinate system X_j).

We calculate [215] the local kinetic energy of turbulized fluid using representation (2.3.2) and (2.3.3):

$$\frac{1}{2}\rho u_i u_i = \frac{1}{2}\rho U_i U_i + \rho U_i \left(\frac{\partial U_i}{\partial X_k}\xi_k + w_i\right) +$$

$$\frac{1}{2}\rho \overline{w}_i \overline{w}_i + \frac{1}{2}\rho \frac{\partial U_i}{\partial X_k}\frac{\partial U_i}{\partial X_m}\xi_k \xi_m +$$

$$\frac{1}{2}\rho \frac{\partial \overline{w}_i}{\partial \xi_k}\frac{\partial \overline{w}_i}{\partial \xi_n}\zeta_k \zeta_n + \rho \frac{\partial U_i}{\partial X_k}\frac{\partial \overline{w}_i}{\partial \xi_m}\xi_k \zeta_m + \tag{3.1.1}$$

$$\rho \overline{w}_i \frac{\partial U_i}{\partial X_k}\xi_k + \rho w_i \frac{\partial \overline{w}_i}{\partial \xi_k}\zeta_k$$

Averaging of expression (3.1.1) over the volume of the mole of λ scale and then over all such moles in the volume ΔV yields the expression for kinetic energy contained in this volume:

$$< \frac{1}{2}\rho u_i u_i >= \frac{1}{2} < \rho > U_i U_i + \frac{1}{2}\rho I_{km}\frac{\partial U_i}{\partial X_k}\frac{\partial U_i}{\partial X_m} + \tag{3.1.2}$$

$$+\frac{1}{2} < \rho \overline{w_i} \overline{w_i} > +2 < \rho i_{km} \Phi_{im} \Phi_{ik} >$$

So, the total energy includes the distortion energy of the velocity field besides the energy of mean and pulsations translation velocity.

The last term at the right-hand part of (3.1.2), that is, the mole distortion energy, can be further transformed as follows [222]:

$$< i_{km}\Phi_{im}\Phi_{ik} > = J_{km} < \Phi_{im} >< \Phi_{im} > + < i_{km}\Phi'_{im}\Phi'_{ik} >$$

$$+[< J'_{km}\Phi'_{im} > + < J'_{km}\Phi'_{ik} >] < \Phi_{im} >; \tag{3.1.3}$$

$$< \Phi_{im} > = \frac{1}{2}(\frac{\partial U_i}{\partial X_m} + \omega_{im})$$

Using this result, we find that the kinetic energy of a turbulent field can be represented as a sum of energies of mean translation motion, mean distortion rate and inner *energy of turbulent chaos* ε:

$$< \frac{1}{2}\rho u_i u_i > = \frac{1}{2} < \rho > U_i U_i + \frac{1}{2} < \rho > I_{km}\frac{\partial U_i}{\partial X_k}\frac{\partial U_i}{\partial X_m} +$$

$$+\frac{1}{2} < \rho > J_{km}(\Omega_{im} + \omega_{im})(\Omega_{ik} + \omega_{ik}) + < \rho > \varepsilon; \tag{3.1.4}$$

The turbulent chaos energy ε is created [215] by pulsations of two types:

$$< \rho > \varepsilon = < \rho > (\varepsilon_w + \varepsilon_\omega)$$

The first corresponds to the translation pulsations:

$$< \rho > \varepsilon_w = \frac{1}{2} < \rho \overline{w_i} \overline{w_i} >;$$

The second is determined by fluctuations of distortion rates:

$$\varepsilon_{\omega} = 2J_{km} < \Phi'_{im} \Phi'_{ik} > +$$

$$2 < i'_{km} \Phi'_{im} >< \Phi_{ik} > + 2 < i'_{km} \Phi'_{ik} >< \Phi_{im} >$$

The kinetic energy of the pulsation field can be determined as

$$\left(\frac{1}{2} \rho u_i u_i\right)' = < \rho > U_i w_i +$$

$$< \rho > (\Omega_{im} + \omega_{im}) (2i_{km} \Phi_{km})' + (\rho\varepsilon)' \tag{3.1.5}$$

The variable $(\rho\mathcal{E})'$ has to be selected in a manner which permits us to keep the sum of expressions (3.1.4) and (3.1.5) equal to initial expression (3.1.1).

We shall use the following expressions, accounting for the previously introduced expressions:

$$U_i < \rho w_i w_j >_J = - U_i R_{ij};$$

$$(\Omega_{pm} + \omega_{pm}) < 2 i_{km} \Phi_{pk} w_j >_j \approx$$

$$(\Omega_i + \omega_i) < 2 \varepsilon_{mpi} i_{km} \Phi_{pk} w_j >_j = - (\Omega_i + \omega_i) \mu_{ij}; \tag{3.1.6}$$

$$< \rho \mathcal{E} u_j > = < \rho > < \mathcal{E} > U_j + < (\rho\mathcal{E})' w_j >_j;$$

$$\varepsilon_{ikl} < \rho u_l u_k >_k = - \varepsilon_{ikl} R_{lk};$$

$$< \sigma_{ij} u_i >_j = < \sigma_{ij} >_j U_i + < \sigma'_{ij} w_j >_j;$$

$$< \varepsilon_{ilk} \xi_l F_k > = C_i ;$$

$$< F_k u_k > \approx < F_k > U_k + < F'_k w_k > + < 2F'_k \Phi_{kn} \xi_n >;$$

$$W = < 2 F'_k \Phi_{kn} \xi_n > - C_i (\Omega_i + \omega_i)$$

We transform the *flux of kinetic energy* analogously to impulse and kinetic moment fluxes. Then

$$< \frac{1}{2} \rho u_i u_i u_j >_j \approx < \frac{1}{2} \rho u_i u_i > U_j + < (\frac{1}{2} \rho u_i u_i)' w_j >_j$$

$$= \frac{1}{2} (< \rho > U_i U_i U_j + U_i < \rho \overline{w}_i \overline{w}_j >_j + U_j < \rho \overline{w}_i \overline{w}_i >_j) \quad (3.1.7)$$

$$+ < 2\rho \Phi_{im} >< \rho i_{km} (\Omega_{ik} + \omega_{ik}) w_j >_j + < (\rho \varepsilon)' w_j >$$

Now the total energy balance (2.2.6), including the mean value of internal energy $<E>$, has the form:

$$< \rho > (\frac{\partial}{\partial t} + U_j \frac{\partial}{\partial X_j})[< E > + \frac{1}{2} U_i U_i +$$

$$\frac{1}{2} I_{km} \frac{\partial U_i}{\partial X_k} \frac{\partial U_i}{\partial X_m} + \frac{1}{2} J_{km} (\Omega_{im} + \omega_{im})(\Omega_{ik} + \omega_{ik})] = \quad (3.1.8)$$

$$\frac{\partial}{\partial X_j} [U_i R_{ij} + (\Omega_i + \omega_i) \mu_{ij} + < \sigma_{ij} >_j U_i] + W +$$

$$< F_i > U_i + C_i (\Omega_i + \omega_i) + \frac{\partial}{\partial X_j} [< \sigma_{ij} w_j >_j -$$

$$- < (\rho E)' w_j >_j - < (\rho \varepsilon)' w_j >_j - < q_j >_j] + < Q > .$$

We shall multiply the balance of angular momentum of mean field (2.3.14) by Ω_i, balance (2.3.16) for mesovortexes by $(\Omega_i + \omega_i)$ and the inertia moment evolution (2.4.5) and (2.4.7), correspondingly, by

$$\frac{1}{2}\varepsilon_{kli}\varepsilon_{mpi}\Omega_l\Omega_p, \quad \frac{1}{2}\varepsilon_{kli}\varepsilon_{mpi}(\Omega_l + \omega_l)(\Omega_p + \omega_p) \tag{3.1.9}$$

Besides, we multiply (2.2.14) by velocity U_i and obtain the equation of so-called "live" forces.

We separate the Reynolds stresses into symmetrical and asymmetrical parts, that is,

$$R^s_{ij} = (1/2)(R_{ij} + R_{ji}), \quad R^a_{ij} = (1/2)(R_{ij} - R_{ji})$$

The subtraction of the sum of indicated products from total energy balance (3.1.8) gives:

$$< \rho > (\frac{\partial}{\partial t} + U_j \frac{\partial}{\partial X_j})(< E > + \mathcal{E}) = (R^s_{ij} +$$

$$\sigma^s_{ij})\frac{1}{2}(\frac{\partial U_i}{\partial X_j} + \frac{\partial U_j}{\partial X_i}) - (R^a_{ij} + \sigma^a_{ij})\,\varepsilon_{ijk}\,\omega_k + < Q > \tag{3.1.10}$$

$$+ \mu_{ij}\frac{\partial(\Omega_i + \omega_i)}{\partial X_j} + \frac{\partial}{\partial X_j}\Big[- < (\rho E)' w_j >_j +$$

$$W - < (\rho\varepsilon)' w_j >_j + < \sigma_{ij} w_j >_j - < q_j >_j\Big]$$

Let us average the entropy balance over the volume ΔV. In the microvolume dv, this equation has the conventional form:

$$\frac{\partial}{\partial t}\rho s + \frac{\partial}{\partial x_j}\rho s u_j = \Xi - \frac{\partial}{\partial x_j}\frac{q_j}{T} + \frac{Q}{T},$$

$$\Xi = \frac{1}{T}(\sigma_{ij} + p\,\delta_{ij})\frac{\partial u_i}{\partial x_j} - \frac{q_j}{T^2}\frac{\partial T}{\partial x_j}, \tag{3.1.11}$$

Here Ξ is local production of entropy, q_j/T is the flux of entropy, T is the temperature. The result is the following one:

$$<\rho>(\frac{\partial}{\partial t} <s> + U_j\frac{\partial}{\partial X_j} <s>) = <\Xi> +$$

$$<\frac{Q}{T}> - \frac{\partial}{\partial X_j}\left[\frac{<q_j>_j}{T} + <(\rho s)'\,w_j>_j\right] \tag{3.1.12}$$

$$<\Xi> = \frac{1}{T}(<\sigma_{ij}>_j + <p>\,\delta_{ij})\frac{\partial U_i}{\partial X_j} +$$

$$<\frac{q_j}{T^2}\frac{\partial T}{\partial x_j}> + \frac{\varphi}{<T>}, \tag{3.1.13}$$

Here the turbulent diffusion type macrotransfer of the entropy appears (at the microscale it corresponds to the convection entropy flux):

$$<(\rho\ s)'\,w_j>_j.$$

The source φ corresponds to the true viscous dissipation of mechanical energy, that is, to its direct transformation into heat at fluctuations of the distortion velocity field ($\sigma'_{ij} = \sigma_{ij} + p\,\delta_{ij}$ is deviator of viscous stress tensor, p – pressure):

$$\varphi = < (\sigma'_{i'j} - < \sigma_{i'j} >_j)(\partial w_i / \partial x_j) > . \tag{3.1.14}$$

If we average the true internal energy E balance equation over the volume ΔV, that is, the heat balance equation, valid in microvolume:

$$\frac{\partial \rho E}{\partial t} + \frac{\partial \rho E u_j}{\partial x_j} = \sigma_{ij} \frac{\partial u_i}{\partial x_j} - \frac{\partial q_j}{\partial x_j} + Q \tag{3.1.15}$$

then we obtain

$$< \rho > (\frac{\partial < E >}{\partial t} + U_j \frac{\partial < E >}{\partial X_j}) = < \sigma_{ij} >_j \frac{\partial U_i}{\partial X_j} - \frac{\partial < q_j >_j}{\partial X_j} + < Q > + \frac{\partial}{\partial X_j} < -(\rho\ E)' w_j >_j - < (\sigma_{ij} - < \sigma_{ij} >_j)(\frac{\partial u_i}{\partial x_j} - \frac{\partial U_i}{\partial X_j}) > . \tag{3.1.16}$$

The Gibbs relation for the mean entropy $< s >$ and for mean energy $< E >$ follows as the difference of (3.1.12) and (3.1.16):

$$\frac{d < E >}{dt} = < T > \frac{d < s >}{dt} - p \frac{d}{dt} \frac{1}{< \rho >} \tag{3.1.17}$$

Here the symbol of substational derivative is used: $d / dt = \partial / \partial t + U_j \partial / \partial X_j$.

We assume the equality

$$\frac{\partial}{\partial X_j} < -(\rho E)' w_j >_j + < T > \frac{\partial}{\partial X_j} < (\rho s)' w_j >_j$$
$$+ (p - < p >)(\frac{\partial u_i}{\partial x_i} - \frac{\partial U_i}{\partial X_j}) >= 0. \quad (3.1.18)$$

3.2. CONSTITUTIVE LAWS FOR TURBULENT FLUIDS

Now we see, that the Gibbs relation in form of (3.1.17) for average values of true internal energy and entropy (see [206]), does not single out the parameters, characterizing the internal structure of turbulent continuum.

Therefore, we pay attention to the internal energy ε of turbulent chaos itself. The corresponding equation can be obtained [215] by the subtracting of equation (3.1.15) for "true" internal energy $<E>$ from equation (3.1.10) for total internal energy

$$< \rho > (\frac{\partial}{\partial t} + U_j \frac{\partial}{\partial X_j})\varepsilon = R^s_{ij} \frac{1}{2}(\frac{\partial U_i}{\partial X_j} +$$
$$\frac{\partial U_j}{\partial X_i}) - R^a_{ij} \; \varepsilon_{ijk} \omega_k + \mu_{ij} \frac{\partial(\Omega_i + \omega_i)}{\partial X_j} + \quad (3.2.1)$$
$$\frac{\partial}{\partial X_j} < -(\rho \varepsilon)' w_j >_j - \Psi + W$$

Here $\Psi = \varphi - \partial < \sigma^*_{ij} w_i >_j / \partial X_j$ is the sink of turbulent energy determined entirely by *viscous turbulent* forces.

The work of turbulent stresses R_{ij}, μ_{ij} at the average translation and angular velocities transforms the mechanical energy of the average field $\partial U_i / \partial X_j$ into the energy of turbulent chaos. The chaos has a "heat" type in the scale of ΔV but in the scale of dv it belongs to the pure mechanical type.

This is in accordance with the Lorentz strange attractor effect where the each small step is deterministic but in larger scale the motion is chaotic [170].

Correspondingly, let us introduce phenomenological turbulent entropy $\in$ and the turbulence temperature θ as follows [215, 218]:

$$< \rho > \theta \frac{d \in}{d t} + \Psi = (R^{s}_{ij} - \frac{1}{3} R_{kl} \delta_{kl} \delta_{ij}) \times$$

$$\frac{1}{2} (\frac{\partial U_i}{\partial X_j} + \frac{\partial U_j}{\partial X_i}) - R^{a}_{ij} \varepsilon_{ijk} \omega_k + \qquad (3.2.2)$$

$$\mu_{ij} \frac{\partial (\Omega_i + \omega_i)}{\partial X_j} + \frac{\partial}{\partial X_j} < - (\rho \varepsilon)^{\prime} w_j >_j$$

or

$$< \rho > \frac{d \in}{d t} + \frac{\Psi}{\theta} = \Xi \qquad (3.2.3)$$

Here Ξ_t is a local production of turbulent entropy, Ψ/θ is a sink of turbulent entropy $\in$.

The separate Gibbs relation is also correct for a turbulent chaos where turbulent pressure $P = - R_{ij}\delta_{ij} / 3$ appears:

$$\frac{d\varepsilon}{dt} = \theta \frac{d \in}{dt} - P \frac{d}{dt} \frac{1}{< \rho >} \qquad (3.2.4)$$

Conventionally, the Gibbs relation (see Section 1.1) is used for determination of state parameters. So, J. Marshall and P. Naghdi [177 – 179] discussed the possible role of turbulent temperature.

However, the introduction of two entropies $\in$ and $< s >$, determined in the macroscale, corresponds to turbulent continuum representation [215] as a thermodynamic complex, composed of two chaos subsystems - turbulent chaos and molecular chaos.

In other words, the work of the Reynolds and couple turbulent stresses as well as turbulent dispersion of internal energy yield the increase of entropy (chaos) of turbulence, but the viscous dissipating stresses decrease the turbulent entropy (turbulent chaos).

The intensity of the sink Ψ/θ of turbulent entropy $\in$ may be also called the *negentropy* source (if one would use the Shroedinger concept developed for biological systems [263]).

In accordance with Richardson's [247] and Kolmogorov's [138-140] idea, *the energy flux $d\mathcal{E}/dt$* along the turbulent eddies hierarchy towards the molecular level appears to *be the parameter in* such a *system that is thermodynamically open.* This was used in many studies, see [138-140, 153, 195], etc.

The energy flux has the following dimension: cm^2 / s^3 [214] and the turbulent field variables can be estimated as follows:

$$w \sim (d\varepsilon/dt)^{1/3} \lambda^{1/3},$$

$$\omega \sim w/\lambda \sim (d\mathcal{E}/dt)^{1/3} \lambda^{-2/3}, \tag{3.2.5}$$

Then we have the following estimates for the specific angular momentum M_λ and the energy $\rho\varepsilon_\lambda$ of the individual turbulent eddy:

$$M_\lambda \sim \rho \lambda^2 \omega \sim \rho \lambda^{4/3} (d\mathcal{E}/dt)^{1/3},$$

$$\rho\varepsilon_\lambda \sim \rho \lambda^2 \omega^2 \sim \rho \lambda^{2/3} (d\mathcal{E}/dt)^{2/3} \sim \rho w^2 \tag{3.2.6}$$

If both, ε_λ and $d\mathcal{E}/dt$, are state parameters, the eddy hierarchy is in a *dynamic equilibrium.* Then we have the estimation of the turbulent eddy length scale:

$$\lambda = (J/\rho)^{1/2} \sim \varepsilon_\lambda^{3/2} / (d\mathcal{E}/dt) \tag{3.2.7}$$

as well as for the eddy cascade:

$$\frac{\omega_i}{\omega_{i+1}} \sim \left(\frac{\lambda_{i+1}}{\lambda_i}\right)^{2/3}, \qquad \frac{M_i}{M_{i+1}} \sim \left(\frac{\lambda_i}{\lambda_{i+1}}\right)^{4/3}$$

$$\frac{\varepsilon_i}{\varepsilon_{i+1}} \sim \left(\frac{\lambda i}{\lambda_{i+1}}\right)^{2/3} \tag{3.2.8}$$

The smallest dissipating eddy will be controlled by molecular viscosity ν, that is,

$$\lambda_\nu \sim \left(\frac{\nu^3}{d\varepsilon_\lambda / dt}\right)^{1/4}, \qquad \omega_\nu \sim \left(\frac{d\varepsilon / dt}{\nu}\right)^{1/2}, \tag{3.2.9}$$

$$M_\nu \sim \rho \nu$$

The dynamic equilibrium corresponds to the *local stationary case*, when the *dissipation* (the energy flux) and the internal *turbulent energy* can also serve as *state parameters*. To prove this point, let us recall the paper [175] by M. Lurie and N. Dmitriev where the following good correlation of turbulent viscosity ν^t, determined by the experimental data by G. Comte-Bello [48], was found:

$$\nu^t \sim \varepsilon / (e_{ij}\, e_{ij})^{1/2}. \tag{3.2.10}$$

with the coefficient depending of the Reynolds number.

In the dynamical equilibrium state equation (3.2.3) simplifies and has the form:

$$< \rho > \frac{d \in}{d t} = \frac{D - \Psi}{\theta} \tag{3.2.11}$$

Here the turbulent dissipation $D = \theta\, \Xi$ of mechanical energy of an average flow is introduced and the turbulence-diffusion of entropy is neglected.

Formula (3.2.11) obviously shows that the entropy $\in$ of a subsystem of turbulent chaos is increases in the case when the turbulent dissipation exceeds the energy sink Ψ into the subsystem of molecular chaos, caused by the viscous damping of turbulent fluctuations. Otherwise, the entropy S may decrease and this is the typical feature of a turbulent subsystem as thermodynamically open.

In a local stationary state $d \in /dt = 0$ that means the exact equality of dissipation intensity to a sink of turbulent energy onto the molecular level [218]

$$D = \Psi = const \tag{3.2.12}$$

The energy flux is realized by dissipation D_e of the average velocity field, governed by conventional turbulent viscosity, as well as by D_ω of the spin of individual turbulent eddies to the turbulent chaos and then by D_ν - to the molecular chaos (the heat). If the corresponding expressions for such dissipation parts are equivalent (see [118, 119, 218]) as parameters of turbulent nonlinear viscosity, it means the steady energy flux and the stationary energy hierarchy.

As it was assumed in [227, 266], the turbulent dissipation D may serve as the state parameter but only in the case of steady energy flux (local equilibrium state).

In this case, we can find the constitutive laws using the Onsager rules (as in Section 1.1) and the turbulent entropy production expression:

$$\Xi = \frac{\mu_{ij}}{\theta}\frac{\partial(\Omega_i + \omega_i)}{\partial X_j} - \frac{1}{\theta} R^a_{ij}\,\varepsilon_{ijk}\,\omega_k + \tag{3.2.13}$$

$$\frac{1}{2\theta}\left(R^s_{ij} - \frac{1}{3}R_{kl}\,\delta_{kl}\,\delta_{ij}\right)\left(\frac{\partial U_i}{\partial X_j} + \frac{\partial U_j}{\partial X_i}\right) +$$

$$\frac{\partial}{\partial X_j}\left(\frac{< -(\rho\,\varepsilon)'\,w_j >_j}{\theta}\right)$$

In this manner we obtain the completing laws, typical [2, 269] for asymmetrical hydrodynamics (or for the Cosserat hydrodynamics):

$$R_{ij}^{s} - \frac{1}{3} R_{kl}\, \delta_{kl}\, \delta_{ij} = \rho\, \nu^{t}{}_{ijmn} \left(\frac{\partial U_m}{\partial X_n} + \frac{\partial U_n}{\partial X_m} \right),$$

$$R_{ij}^{a} = 2\rho\, \gamma_{ijnm}\, \varepsilon_{nmk}\, \omega_k \,, \qquad (3.2.14)$$

$$\mu_{ij} = 2\rho\, \eta_{ijkmln}\, J_{\ln} \frac{\partial(\Omega_k + \omega_k)}{\partial X_m},$$

$$q_j \equiv \langle -(\rho\varepsilon)'\, w_j \rangle_j = \kappa_{ji} \frac{\partial}{\partial X_i} \left(\frac{1}{\theta} \right),$$

The last expression (3.2.14) means that the flux of turbulent "heat" is proportional to gradient of *turbulent "temperature"* [215]. The tensor turbulent coefficients of viscosity and of heat conductivity themselves are functions of changing mesostructure, that is, of average turbulent parameters[12]. They are the Onsager coefficients $L_{\beta\alpha}$ and, in accordance with (3.2.12) are functions of dissipation D of the mean turbulent field energy.

This reflects the turbulent field dependency on its thermodynamic fluxes Y_β and means nonlinearity of closing laws (3.2.14) and that an exclusion of one (or of part) of the forces X_α (of fluxes Y_β) can change the Onsager coefficient matrix $L_{\alpha\beta}$. Such a situation is typical for self-organizing (synergetic) systems [103].

After account of this is taken, the usual requirement of positive determination of every term in the products sum $\Sigma = L_{\alpha\beta} X_\alpha X_\beta$ becomes not obligatory. The only necessity is that the Σ value (or the total dissipation D) should be strictly positive. It follows from here that, in principle, the superposition of different fluxes can lead to *negative* values of some elements of matrix $L_{\alpha\beta}$.

Moreover, in locally non-equilibrium cases the turbulent entropy may decrease but this situation has to be described by a kinetic consideration. For example, the adequate descriptions of larger eddies generation by the energy

---

[12] Some more complicated constructions of the constitutive laws can be found in the literature. The turbulence of quickly changing turbulent structures (non-local relations) was considered [215, 222, 282]. The account for anisotropy should be carried out within corresponding rules [175, 259, 295, 306].

influx from smaller ones have to be developed. This problem is essential for explanation of a number of geophysical phenomena [29, 236, 247, 268, 300] and can be solved by choice of a proper combination of kinetic parameters (compare [26]).

The known conclusion about *negative turbulent viscosity* for some flows is based just on the comparison of average fields of U_i and R^s_{ij} (see [110, 268]) within the limits of conventional approach, that is, without accounting for the Reynolds stresses asymmetry, etc. However, this does not exclude the possibility of true negative values of some kinetic coefficients.

3.3. KINETIC ANALYSIS OF TURBULENCE

Let us try to find the structure of expression (3.2.14) in a plane case. The constitutive laws give (*P* is a turbulent pressure):

$$R_{12} + R_{21} = 2\rho\nu^t\left(\frac{\partial U_1}{\partial X_2} + \frac{\partial U_2}{\partial X_1}\right)$$

$$R_{11} = -R_{22} = -P + 2\rho\nu^t\frac{\partial U_1}{\partial X_1}, \tag{3.3.1}$$

$$R_{12} - R_{21} = 4\rho\gamma^t\omega_3, \qquad \omega_3 \equiv \omega$$

We can rewrite the Onsager expression (3.2.14) for couple stresses in the form of "turbulent diffusion" of an angular momentum:

$$\mu_{3\,j} = < -M'_3 w_j >_j = 2\eta\frac{\partial M_3}{\partial X_j} = 2\rho\eta J\frac{\partial(\Omega_3 + \omega_3)}{\partial X_j} +$$

$$+ 2\rho\eta(\Omega_3 + \omega_3)\frac{\partial J}{\partial X_j}, \qquad j = 1, 2, \qquad \Omega \equiv \Omega_3 \tag{3.3.2}$$

Correspondingly, we have to account the evolution of the moment of inertia of a mesovortex:

$$\Pi_j \equiv - < \rho J' w_j >_j = 2\rho \varsigma \frac{\partial J}{\partial X_j}, \quad \varsigma = \varsigma^* |\omega|$$

This result appears to be somewhat more general, than the analysis carried out above (relations (3.3.2) coincide with (2.3.17), (2.4.9) just when $J = const$). Here η, ζ, ζ^* are kinetic coefficients.

Let us carry out, for example, the analysis of kinetics of turbulent intermixture of impulse and of angular momentum. We consider the following components on the base of kinetics of an "average" mole motion inside the turbulent flow accounting for its "average" pulsation:

$$R_{12} = -\rho < w_1 w_2 >_2 ,$$
$$\mu_{31} = -\rho < (i\Phi)' w_2 >_2 . \tag{3.3.3}$$

The estimation of pulsation velocities w_1, w_2 will be performed as differences of mean velocities $U = U_1$ in adjacent flow layers, separated by a small distance λ.

Such an estimation [96] is explained by the possibility of the identification of λ with mean mixing length l, along which the fluid mole preserves [169, 240] its impulse. The mole, migrating between the indicated layers, creates the velocity pulsation because its velocity differs from the velocities of particles - aborigines.

If there is $U_1 >> U_2$, our estimations should be done for the component U_1. Besides, the isotropy of absolute values of pulsation $| w_1 | \sim | w_2 |$ is accepted as an approximation.

Correspondingly, some impulse transfer is caused by the difference of mean translation velocities of moles, separated by "mixing" length:

$$X_2 = const, \qquad X_2 + l = const.$$

Let the fluid mole exchange take place at the boundary of the layer $X_2 = const$. The impulse difference Δ_U of particles, coming in and coming out, is as follows:

$$\Delta_U(\rho w_1) = \rho U_1(X_2 + l) - \rho U_1(X_2) =$$

$$= \rho\, l\, (\partial U_1 / \partial X_2), \tag{3.3.4}$$

Due to the isotropy we obtain if $\rho = const$:

$$|\Delta_U w_2| \sim |\Delta_U w_1| \sim l\, |\partial U_1 / \partial X_2|. \tag{3.3.5}$$

The "average" micromotion simultaneously takes place inside the "average" field of spin (its own angular) velocity $\phi = -\omega / 2$. We shall accordingly assume that the particle-migrant, accomplishing the λ - length run, passes through the plane mesovortex street of $\kappa = \lambda\omega$ intensity[13]. Then the particles, crossing the street in the opposite direction, will carry the additional impulse proportional to the same street intensity [213]. As a result we obtain:

$$\Delta_\omega(\rho w_1) = \rho\lambda\omega - (-\rho\lambda\omega) = 2\rho\lambda\omega. \tag{3.3.6}$$

It is essentially that this mesovortex street presence affects only tangential (in respect of the street) velocity components but its normal components do not change. Therefore,

$$\rho w_1 = \rho\, l(\partial U_1 / \partial X_2) + 2\rho\lambda\omega$$

$$|w_2| = l\, |\partial U_1 / \partial X_2|. \tag{3.3.7}$$

[13] In the famous textbook by N. Kochin et al [137] the vortex layer is characterized by the intensity that can be represented as $\kappa = \lambda\Gamma$, where Γ is velocity circulation around an individual vortex line, a number of which composes the layer.

If we use now estimations (3.3.5), accounting [258] for the choice of sign for ω_2 (the stress must have the same sign with transported variable), then in accordance with (3.3.3), we obtain

$$R_{12} = \rho l \mid \partial U_1 / \partial X_2 \mid (l \partial U_1 / \partial X_2 + 2\lambda\omega). \tag{3.3.8}$$

The comparison with (3.3.4) gives the estimations of turbulent shear ν^t and rotary γ^t turbulent viscosity:

$$\nu^t = l^2 \mid \partial U_1 / \partial X_2 \mid, \qquad \gamma^t = l\lambda \mid \partial U_1 / \partial X_2 \mid \tag{3.3.9}$$

If we evaluate the R_{21} component, then the motion of particles through the layer $X_1 = const$ should be examined, the mesovortex street [213] being oriented in another way. Therefore, we have

$$\mid w_1 \mid = l \mid \partial U_1 / \partial X_2 \mid,$$

$$\rho w_2 = \rho l (\partial U_1 / \partial X_2) - 2\rho \lambda \omega, \tag{3.3.10}$$

$$R_{21} = \rho l \mid \partial U_1 / \partial X_2 \mid (l \partial U_1 / \partial X_2 - 2\lambda \omega).$$

Let us calculate the transport pulsation of the angular momentum:

$$\Phi' = l \frac{\partial (-\partial U_1 / \partial X_2 + \omega)}{\partial X_2}, \quad i' = l_J \frac{\partial J}{\partial X_2},$$

$$(i\Phi)' = Jl \frac{\partial (-\partial U_1 / \partial X_2 + \omega)}{\partial X_2} + \tag{3.3.11}$$

$$(-\partial U_1 / \partial X_2 + \omega) l_J \frac{\partial J}{\partial X_2}.$$

Assuming the analogous argumentation about the sign choice as it was done above, we obtain the expression for the couple stresses:

$$\mu_{3i} = J\, l^2 |\omega| \frac{\partial(-\partial U_1 / \partial X_2 + \omega)}{\partial X_2} +$$
$$l\, l_J (-\partial U_1 / \partial X_2 + \omega) \mid \omega \mid \frac{\partial J}{\partial X_2} \tag{3.3.12}$$

Hence, we have the following expressions for the gradient-eddy viscosity η and for coefficient of intermixture ζ for the inertia moment:

$$\eta = \frac{1}{2} J\, l^2 \mid \omega \mid, \qquad \zeta = \frac{1}{2} l\, l_J (-\partial U_1 / \partial X_2 + \omega) \mid \omega \mid \tag{3.3.13}$$

Now let us recall that geometrical moment of inertia J is equal to the average mole inertia moment divided by its volume (to its area in a plane case). This gives the estimation $J = (1/2)\, r^2$ where r is the average radius of a mole.

If the internal structure of the turbulent continuum is characterized only by length scale λ, then $r \sim \lambda$ and equation (2.4.16) describes the evolution equation of the Prandtl's *mixing length* $l\,(=\lambda)$.

That is, the evolution of the geometrical moment coincides with the evolution of the (square of) mixing length of the Prandtl mole:

$$\frac{\partial \lambda^2}{\partial t} + U_j \frac{\partial \lambda^2}{\partial X_j} = \frac{\partial}{\partial X_j} (\zeta \frac{\partial \lambda^2}{\partial X_j}) - Q_\lambda, \tag{3.3.14}$$

Here the following sink of internal turbulent structure appears:

$$Q_\lambda = < -\Phi' i' > \tag{3.3.15}$$

We have to recall that in the continuum mechanics all phenomenological coefficients can depend only on variables that are invariant at rigid translation and rotation of the coordinate system.

Therefore, variable gradients ($\partial U_i/\partial X_j$), strain rates ($e_{ij}$) and spin velocity ($\omega_j$) can be used to reflect the nonlinear properties but not the average rotation (Ω_j).

The Taylor theory of eddy transfer [96, 278] is well known in literature. It was introduced as an alternative to Prandtl's theory of impulse transfer [239]. The latter included impulse balance (2.4.15) and constitutive law (3.3.8) when $\omega = 0$, or, in other words, the conventional hypothesis of the symmetry of the Reynolds stress tensor ($R_{12} = R_{21}$) was accepted. We recall that it is just a special form of the angular momentum balance.

The Taylor equation was developed on the basis of a kinetic analysis of mixing the process of vorticity in a two-dimensional velocity field. Let us consider the corresponding vorticity balance that is valid at microscale level. It has a divergent form:

$$\frac{\partial \Phi}{\partial t} + \frac{\partial \Phi u_j}{\partial x_j} = \nu \frac{\partial^2 \Phi}{\partial x_j \partial x_j} \tag{3.3.16}$$

Its spatial averaging, i.e. the averaging over the cross-section of eddy tubes with an account for rule (2.1.4), leads [11], to equation

$$\frac{\partial < \Phi >}{\partial t} + U_j \frac{\partial < \Phi >_j}{\partial X_j} = \frac{\partial}{\partial X_j} < - \Phi' w_j >_j + \nu \frac{\partial}{\partial X_j} < \frac{\partial \Phi}{\partial x_j} >_j \tag{3.3.17}$$

Now according to Section 2.1 we can introduce average and spin velocities: $<\Phi> = (\Omega + \omega)/2$. Then we obtain:

$$\frac{\partial (\Omega + \omega)}{\partial t} + U_j \frac{\partial (\Omega + \omega)}{\partial X_j} = \frac{\partial}{\partial X_j} \eta \frac{\partial (\Omega + \omega)}{\partial X_j} + \frac{\partial \varpi_j}{\partial X_j}; \tag{3.3.18}$$

$$< - \Phi' w_j >_j = \eta \frac{\partial (\Omega + \omega)}{\partial X_j};$$

$$\eta = \lambda^2 |\omega|;$$

$$\varpi_j = [(\Omega + \omega) - 2 < \Phi >_j] U_j + < \nu \frac{\partial \Phi}{\partial x_j} >_j$$

The equation of the eddy transport in a turbulent flow follows from (3.3.18) if we neglect the ϖ_j - flux.

On the other hand the equation of angular momentum balance (2.3.9) and inertia moment evolution (2.4.8) yield *the eddy transfer in the form of (3.3.14) if one assumes* $\zeta = -\eta$, $\eta > 0$. It appears that *the eddy's moment of inertia (and its sizes) grows as the eddy's diffusion is developing due to its negative viscosity.* The conclusion is in accordance with the growth of the vortex inertia moment due to molecular viscosity (see Chapter 1). This means also that the Prandtl mixing length is growing in time.

It is impossible now to oppose the Prandtl transport theory and the Taylor theory. They are both the essential consequence of the analysis carried out here if we account for the spin velocity in accordance with the spatial averaging. Both equations are necessary for the solutions, because of the real existence of additional degree of freedom ω_i (which is invariant to rigid rotation).

The situation is similar to the case when the spatial averaging is applied to the *helicity* equation that was obtained by multiplication of vortex evolution equation (1.1.14) by local velocity [188 - 190].

3.4. TURBULENCE PROBLEM FORMULATION

Let us illustrate the suggested turbulence theory by some examples. They will show possible boundary conditions as well as limits of the theory in its linear variant.

As it was mentioned above, the simplest, theoretically motivated hypothesis consists in the dependence of the Onsager coefficients on the dissipation D. Particularly, the turbulent shear viscosity ν^t depends on D.

If the dissipation effects of rotation are negligible, in the case of isotropy and incompressibility we have that $\nu^t = \nu^t (D)$, where D is the dissipation of the mean velocity field:

$$D = \rho \nu^t \frac{1}{4}\left(\frac{\partial U_i}{\partial X_j} + \frac{\partial U_j}{\partial X_i}\right)\left(\frac{\partial U_i}{\partial X_j} + \frac{\partial U_j}{\partial X_i}\right) \equiv \rho \nu^t e_{ij} e_{ij} \tag{3.4.1}$$

The nonlinear dependency of turbulent viscosity is clearly shown by (3.2.6). On the base of the dimension analysis it is possible to suggest:

$$\nu^t = \nu \; f(N), \qquad N = \frac{\lambda^2}{\nu} \left| e_{ij} \right| \tag{3.4.2}$$

Expression (3.4.2) can be simplified if $f(N)$ is a linear function:

$$\nu^t = \lambda^2 \left| e_{ij} \right| \tag{3.4.3}$$

which can be written as

$$\nu^t = 4 \lambda^2 \sqrt{\left(\frac{\partial U_i}{\partial X_j} + \frac{\partial U_j}{\partial X_i}\right)\left(\frac{\partial U_i}{\partial X_j} + \frac{\partial U_j}{\partial X_i}\right)}, \tag{3.4.4}$$

This is a generalization of the famous Prandtl's formula [96, 240]

$$\nu^t = \lambda^2 \left| \frac{\partial U_1}{\partial X_2} \right| \tag{3.4.5}$$

which is valid for the *plane-parallel flow:*

$$U_1 = U_1 (X_2), \; U_2 = 0 . \tag{3.4.6}$$

In other words, the thermodynamic analysis, carried out above, includes classical versions of the turbulence theory. Moreover, generalization of turbulent viscosity (3.4.2) can be applied for more complicated flows than the plane-parallel flow.

For example, let us consider a plane problem of a *free stream boundary* (Figure 3.4.1) when the velocity components are expressed [68, 96, 258] by the stream function ψ, that is,

$$U_1 = \frac{\partial \psi}{\partial X_2}, \quad U_2 = -\frac{\partial \psi}{\partial X_1}. \tag{3.4.7}$$

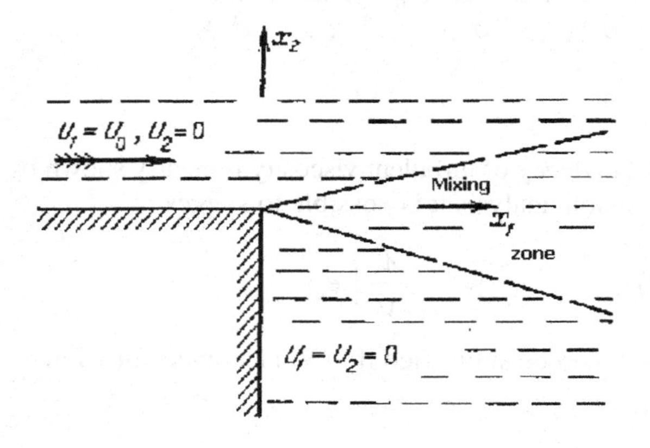

Figure 3.4.1. Scheme of turbulent flow at the corner

Assume that it has the following form:

$$\psi = U_0 X_1 F(\eta), \quad \eta = \frac{X_2}{X_1} \tag{3.4.8}$$

Here U_0 is the velocity of the ambient flow. The incompressibility and the Reynolds equations of impulse balance:

$$\frac{\partial U_1}{\partial X_1} + \frac{\partial U_2}{\partial X_2} = 0,$$

$$U_1 \frac{\partial U_1}{\partial X_1} + U_2 \frac{\partial U_1}{\partial X_2} = \frac{1}{\rho} \frac{\partial R_{12}}{\partial X_2}, \tag{3.4.9}$$

$$R_{12} = 2\rho\lambda^2\left[(\frac{\partial U_1}{\partial X_1})^2 + \frac{1}{4}(\frac{\partial U_1}{\partial X_2} + \frac{\partial U_2}{\partial X_1})^2 + (\frac{\partial U_2}{\partial X_2})^2\right]^{\frac{1}{2}} \times$$

$$(\frac{\partial U_1}{\partial X_2} + \frac{\partial U_2}{\partial X_1})$$

will be transformed [218] into the ordinary differential equation:

$$F = \beta^2 \frac{d}{d\eta}(\sqrt{1 + 6\eta^2 + \eta^4}(1 - \eta^2)\frac{d^2 F}{d\eta^2}) \tag{3.4.10}$$

if $\lambda = \beta X_1$, $\beta = const$.

This can be checked by the following representations for variables involved in the analysis:

$$U_1 = U_0 F', \qquad U_2 = U_0(\eta F' - F),$$

$$F' = \frac{dF}{d\eta},$$

$$\frac{\partial U_1}{\partial X_1} = -\frac{\partial U_2}{\partial X_2} = -\frac{U_0 \eta F''}{X_1},$$

$$\frac{\partial U_1}{\partial X_2} = \frac{U_0 \eta F''}{X_1},$$

$$\frac{\partial U_2}{\partial X_1} = -\frac{U_0 \eta^2 F''}{X_1}.$$

The known solution [258], where the expression

$$R_{12} = \rho \lambda^2 \mid \partial U_1 / \partial X_2 \mid (\partial U_1 / \partial X_2), \tag{3.4.11}$$

was used instead of (3.4.7) under the same conditions, yields a simpler differential equation which approximates (3.4.6) when $|\eta| \gg 1$:

$$F = \beta^2 \frac{d^3 F}{d\eta^3} \tag{3.4.12}$$

The comparison of equation (3.4.12) with experiments [258] is quite satisfactory in the range $-0{,}176 \leq \eta \leq 0{,}083$.

Turbulent flows at a rigid wall

Within the one-dimensional flow of turbulent fluid through a tube or a plane channel, the tangential force *at its walls* can be represented as a sum of two terms:

$$\tau^* = \nu^t \left(\frac{\partial U}{\partial n}\right)^* + 2\gamma\omega^*, \tag{3.4.13}$$

The first term is usual for the Boussinesq turbulent viscosity concept and n means the normal to the wall surface. The second term conforms to presence of eddies with spin velocity $\phi = -\omega/2$, generated by wall asperity if the turbulent continuum spreads up to the wall.

Let us describe the latter effect by a constant source of eddies at the wall. These eddies rotate slower than the average vortices, that is [223],

$$\omega = \kappa\, \omega^*, \quad \kappa = -\, sgn\ \Omega = sgn\left(\frac{\partial U}{\partial n}\right), \quad \omega^* > 0, \tag{3.4.14}$$

The scale analysis is shows that

$$\omega^* \frac{a}{U} = \varphi\left(\frac{a}{L}; \frac{Ua}{\nu}\right). \tag{3.4.15}$$

Here a/L is a relative roughness, a is a height of roughness topography,

$$Re = \frac{Ua}{\nu}, \qquad Sl = \omega * \frac{a}{U} \tag{3.4.16}$$

are the corresponding Reynolds and modified *Strukhal* numbers.

In the simplest linear version of dependency $Re\ (Sl)$ we have

$$\omega^* = N(\frac{a}{L})\frac{U^2}{\nu}. \tag{3.4.17}$$

Thus, a spin velocity of eddies generated at the walls is proportional to the square of characteristic flow velocity (to a flow rate).

We have to characterize spin intensity at a wall by the introduction of average velocity gradients into (3.4.15):

$$\omega^* = \alpha \frac{a^2}{\nu} \left|\frac{\partial U}{\partial n}\right|^* \left(\frac{\partial U}{\partial n}\right)^* = -\alpha \frac{a^2}{\nu} |\Omega| * \Omega * \tag{3.4.18}$$

If $\alpha > 0$, the nonlinear slowing of total vorticity $\Omega + \omega$ at a wall is evident[14]. The introduction of (3.4.18) into (3.4.13) gives the Prandtl formula valid at a wall:

$$\frac{\tau *}{\rho} = \nu \left|\frac{\partial U}{\partial n}\right|^* + 2\alpha\gamma \frac{a^2}{\nu} \left|\frac{\partial U}{\partial n}\right|^* \left(\frac{\partial U}{\partial n}\right)^* \tag{3.4.19}$$

This means that the internal length of turbulent mesostructure λ at a wall is determined by the asperity value:

$$\lambda = a\sqrt{\frac{\alpha\gamma}{\nu}}$$

Inside a flow this length will change and therefore the inertia moment will change also: $J_{km} = J\ \delta_{km}$, $J \approx \lambda^2$. Let us consider what nonlinearity (3.4.19) can give if inside a flow the linear variant of the theory is valid.

---

[14]Note that the conditions $\Phi = 0$ or, more generally, $\omega = -(1-\alpha)\Omega$, $0 \le \alpha \le 1$ were used [1, 2, 128, 132 – 136, 186, 235] in the asymmetric (polar) hydrodynamics.

The turbulent stream along an infinite plate at zero pressure gradient ($\partial P / \partial x = 0$) when $U = U(y)$, $\omega = \omega(y)$, $\Phi = -(1/2)(\partial U / \partial y)$ is governed by the following equations:

$$\frac{d}{dy}\left(\nu \frac{dU}{dy} + 2\gamma\omega\right) = 0 \tag{3.4.20}$$

$$\frac{d}{dy}\left[\varsigma\left(-\frac{dU}{dy} + \omega\right)\frac{dJ}{dy} + 2\eta J \frac{d}{dy}\left(-\frac{dU}{dy} + \omega\right)\right] = 4\gamma\omega \tag{3.4.21}$$

$$\frac{d}{dy}\left(\xi \frac{dJ}{dy}\right) = 0 \tag{3.4.22}$$

The solution of (3.4.22) determines the linear spreading of the eddy at a distance from a wall:

$$J(y) = A_J y + J_0$$

The next step is exclusion of U from (3.4.21) that is transforming into the Bessel equation:

$$z\frac{d^2\omega}{d\,y^2} + \left(1 + \frac{\varsigma}{2\eta}\right)\frac{d\omega}{d\,y} - \beta^2\omega = 0$$

$$z = \frac{A_J y + J_0}{A_J} \tag{3.4.23}$$

Its solution has the form:

$$\omega = z^{\varphi/2}\left[C_1 I_\varphi(2\beta\sqrt{z}) + C_2 K_\varphi(2\beta\sqrt{z})\right] \tag{3.4.24}$$

$$\beta^2 = \frac{2\nu\gamma}{\eta(\nu - \gamma)A} > 0\,, \qquad \varphi = -\frac{\varsigma}{2\eta}$$

If $A_J = 0$, equation (3.4.23) is simpler:

$$\frac{d^2\omega}{d\,y^2} = k^2\omega\,; \qquad k^2 = \frac{2\nu\gamma}{J_0\eta(\nu - \gamma)} \tag{3.4.25}$$

Then the exponential decay of spin rotation is evident:

$$\omega = \omega_0 \exp(-ky). \tag{3.4.26}$$

More slow eddy attenuation follows asymptotically from (3.4.20) at $C_1 = 0$, $\varphi = 1$, $A \neq 0$:

$$\omega = C_2 \sqrt{\frac{\pi}{2}} \exp(-2\beta\sqrt{z}). \tag{3.4.27}$$

The velocity profile $U(y)$ can be found from the first integral of (3.4.20):

$$\frac{dU}{dy} + 2\frac{\gamma}{\nu}\omega = \frac{\tau^*}{\rho\nu},$$

that is, at $U(0) = 0$ we have:

$$U = \frac{\tau^*}{\rho\nu} y - 2\frac{\gamma}{\nu k}\omega^* [1 - \exp(-ky)] \tag{3.4.28}$$

Close to the wall we have:

$$U \approx \left(\frac{\tau^*}{\rho\nu} - 2\frac{\gamma}{\nu}\omega^*\right) y \tag{3.4.29}$$

As we can see, tangential stress τ satisfies (3.4.19).

Stationary flow in a plane channel

In this case

$$\partial p / \partial x = \rho h = const \neq 0$$

that is, the head h is acting and equations (3.4.20) – (3.4.22) has the form, if $J = J_0$:

$$\frac{d}{dy}\left(\nu \frac{dU}{dy} + 2\gamma\omega\right) = h \tag{3.4.30}$$

$$\frac{d}{dy}\left[\eta J_0 \frac{d}{dy}\left(-\frac{dU}{dy} + \omega\right)\right] = 2\gamma\omega \tag{3.4.31}$$

They give two integrals if the channel width is $2L$:

$$\nu \frac{dU}{dy} + 2\gamma\omega = hy + C \tag{3.4.32}$$

$$\omega = \omega_0 \frac{sh\,(ky)}{sh\,(kH)} \tag{3.4.33}$$

Let solution (3.4.32) - (3.4.33) satisfy (3.4.18), symmetry condition (at $y = 0$) and boundary condition $U(y) = 0$ at $y = \pm H$. Then

$$U = -\frac{H^2}{2\nu}\left(1 - \frac{y^2}{H^2}\right) h +$$

$$2\alpha \frac{\gamma H^2}{\nu^4 k} cth\,(Hk)\left(1 - \frac{ch\,(ky)}{ch\,(kH)}\right) h\,|h| \tag{3.4.34}$$

The flow rate is found by integration across the channel:

$$Q = \int_{-L}^{+L} U dy = -\frac{2}{3}\frac{H^3}{\nu} h\left[1 - |h| \frac{6\alpha\gamma a^2}{\nu^3 k}\left(cth\,kH - \frac{1}{kH}\right)\right] \tag{3.4.35}$$

Resolving (3.4.31) for the head, we can get that $h = AQ + BQ^2$ – typical formula for turbulent flow through channels.

However, the velocity profile does not coincide with the turbulent one and, of course, because of existence of the known *laminar sublayer* [96, 169, 240]. Formally, one can reach agreement with the velocity profile under the constant coefficients of viscosity ν, γ, η just under an introduction of artificial "sliding velocity" U_s at the walls of the channel [200 - 203].

A. Eringen [80, 81] suggested practically the same idea earlier and it was used and developed by G. Ahmadi et al [5] who applied the zero velocity condition at the wall keeping the inertia moment constant. These authors fitted the turbulent viscosity as a function of the Reynolds numbers on the base of experiments by J. Laufer [155, 156] and G. Comte - Bello [48].

CHAPTER 4

TURBULENT WAKES IN THE ATMOSPHERE

4.1. TURBULENT WAKE OF A BODY

The simplest case is when eddy rotation happens inside the plane of an average flow. As it is known, the Karman vortex street [96, 137] appears in a wake behind a body, for example, behind a raised cylindrical chimney. This problem will be considered in the framework of the suggested theory with the boundary condition, formulated by N.E. Zhukovskiy [307] for trailing vortexes.

The body of revolution also generates the turbulent attached jet, considered here. One of the possible applications is the spreading of chemicals behind an airplane. Provision is made in the analysis for suspended particles.

Let us consider turbulent flows in a plane case by the equation for vorticity $\boldsymbol{\Omega}$ of the mean velocity field $\boldsymbol{U}$, as C.C. Lin has done [165]. To develop the solution, Lin had made some assumptions that were typical for the theory of a boundary layer. Besides the known solution

$$\Omega = \frac{1}{2} rotU \sim \frac{\varphi}{x} \exp(-\varphi^2), \qquad \varphi = \frac{y\sqrt{U_\infty}}{2\sqrt{\nu x}}, \tag{4.1.1}$$

where x is a coordinate (macro) along the wake of a body, y is a coordinate across the wake and U_∞ is a velocity of an external flow, Lin shows also the existence of a solution of the type:

$$\Omega \sim \frac{1}{\sqrt{x}} \{ \exp[-(\varphi - \varphi_0)^2] - \exp[-(\varphi + \varphi_0)^2] \}, \tag{4.1.2}$$

$$\varphi_0 = \frac{d\sqrt{U_\infty}}{2\sqrt{\nu x}}$$

Here d is the distance between the points of eddies coming off. The solution (4.1.2) at the point $x = 0$, where the streamlined body is located, corresponds to the dipole modeling the body[15]. Nevertheless, solution (4.1.2) could not explain the known experimental deviations [46, 90] from the self-similar solution (4.1.1).

[15] The oscillating cofactors [165] could model a discrete structure of the Karman eddy wake.

According to the point of view, being developed here, we assume that eddies generated by the body, have their own vorticity differing from the mean value Ω by the spin ω.

Linear approximations of a boundary layer, impulse balance (2.2.14) and angular momentum balance (2.3.9) are the following ones:

$$U_\infty \frac{\partial U^\Delta}{\partial x} = -\frac{1}{y^n}\frac{\partial}{\partial y}(y^n R),$$
$$R = -\nu \frac{\partial U^\Delta}{\partial y} + 2\gamma\omega, \tag{4.1.3}$$

$$U_\infty \frac{\partial}{\partial x}(\Omega + \omega) = 2\eta \frac{\partial}{\partial y}\{\frac{1}{y^n}\frac{\partial}{\partial y}[y^n(\Omega + \omega)]\} - \frac{4\gamma}{J}\omega. \tag{4.1.4}$$

Here $U^\Delta = U - U_\infty$ is a velocity defect, $\rho R = R_{xy} \neq R_{yx}$ is the Reynolds tangential stress. The eddy viscosity coefficients ν, γ, η and the moment of inertia J are supposed to be constant. When $n = 0$, the equations correspond to a plane flow, but when $n = 1$ the equations correspond to wake of body of revolution (in cylindrical coordinate system).

We model the streamlined body as a source of eddies, i.e., as a dipole for a plane problem:

$$\Omega(x = 0) = K[\delta(y + \frac{d}{2}) - \delta(y - \frac{d}{2})], \qquad K > 0$$
$$\omega(x = 0) = -G[\delta(y + \frac{d}{2}) - \delta(y - \frac{d}{2})], \qquad G > 0 \tag{4.1.5}$$

and as an annular source for an axially symmetrical one:

$$\Omega(x = 0) = -K\delta(y - \frac{d}{2}), \qquad \omega(x = 0) = G\delta(y - \frac{d}{2}). \tag{4.1.6}$$

As it is known, the idea about the use of annular eddy sources belongs to N. E. Zhukovskiy [307]. Here δ is the Dirac delta-function.

Far from the body, the boundary conditions have a form:

$$U^{\Delta} = \frac{\partial U^{\Delta}}{\partial y} = 0, \qquad \omega = \frac{\partial \omega}{\partial y} = 0, \qquad |\, y \,| \to \infty \tag{4.1.7}$$

Rewrite equations (4.1.3), (4.1.4) in the following form:

$$U_{\infty} \frac{\partial \Omega}{\partial x} - \nu \frac{\partial}{\partial y}\Big[\frac{1}{y^n}\frac{\partial}{\partial y}(y^n\, \Omega)\Big] = - \gamma \frac{\partial}{\partial y}\Big[\frac{1}{y^n}\frac{\partial}{\partial y}(y^n\, \omega)\Big],$$

$$U_{\infty} \frac{\partial \omega}{\partial x} - (2\eta + \gamma) \frac{\partial}{\partial y}\Big[\frac{1}{y^n}\frac{\partial}{\partial y}(y^n\, \omega)\Big] + \frac{4\gamma}{J}\,\omega = (2\eta - \nu) \frac{\partial}{\partial y}\Big[\frac{1}{y^n}\frac{\partial}{\partial y}(y^n\, \Omega)\Big]. \tag{4.1.8}$$

We suppose the difference of tangential stresses R_{xy} and R_{yx} much smaller than the stresses themselves, i.e. $|\, \nu\, \Omega \,| >> |\, \gamma\, \omega \,|$. Let us require additionally that $|(2\eta + \gamma)\omega| >> |(2\eta - \nu)\Omega|$.

We begin with the system of homogeneous equations

$$U_{\infty} \frac{\partial \Omega_0}{\partial x} - \nu \frac{\partial}{\partial y}\Big[\frac{1}{y^n}\frac{\partial}{\partial y}(y^n\, \Omega_0)\Big] = 0,$$

$$U_{\infty} \frac{\partial \omega_0}{\partial x} - (2\eta + \gamma) \frac{\partial}{\partial y}\Big[\frac{1}{y^n}\frac{\partial}{\partial y}(y^n\, \omega_0)\Big] + \frac{4\gamma}{J}\,\omega_0 = 0. \tag{4.1.9}$$

If the mentioned inequalities are accepted, solutions of homogeneous equations (4.1.9) yield the main terms of Ω and ω. Accounting for this, we shall seek an approximate solution of system (4.1.8) converting it into the following equation system

$$U_\infty \frac{\partial \Omega}{\partial x} - \nu \frac{\partial}{\partial y}\left[\frac{1}{y^n}\frac{\partial}{\partial y}(y^n \Omega)\right] = -\gamma \frac{\partial}{\partial y}\left[\frac{1}{y^n}\frac{\partial}{\partial y}(y^n \omega_0)\right],$$

$$U_\infty \frac{\partial \omega}{\partial x} - (2\eta + \gamma)\frac{\partial}{\partial y}\left[\frac{1}{y^n}\frac{\partial}{\partial y}(y^n \omega)\right] + \frac{4\gamma}{J}\omega = \tag{4.1.10}$$

$$= (2\eta - \nu)\frac{\partial}{\partial y}\left[\frac{1}{y^n}\frac{\partial}{\partial y}(y^n \Omega_0)\right],$$

where Ω_0, ω_0 are the solutions of homogeneous system (4.1.9). One can develop the solutions of (4.1.10) with boundary conditions (4.1.5) - (4.1.7) in the form of integrals [118, 216].

Under the condition that $4\sqrt{\nu x/U_\infty} >> d$ they assume the form

$$\Omega = -\frac{y}{2}\left(K + \frac{\gamma}{2\eta+\gamma}G\right)\frac{d^{1+n} U_\infty^{1+n/2}}{16^n \pi^{(1-n)/2} (x\nu)^{1+n/2}} \exp\left(-\frac{U_\infty y^2}{4\nu x}\right)$$

$$+ \frac{y}{8}\frac{\gamma G d^{1+n} U_\infty^{1+n/2}}{4^n \pi^{(1+n)/2}} \times \tag{4.1.11}$$

$$\int_0^x \exp\left(-\frac{4\gamma t}{J U_\infty} - \frac{U_\infty y^2}{4(\nu x + \beta t)}\right)(\nu x + \beta t)^{-(5+n)/2} \times$$

$$\left(3 - n - \frac{U_\infty y^2}{2(1+n)(\nu x + \beta t)}\right) dt,$$

$$\omega = \frac{y}{2}(G - \frac{\gamma}{\nu}K)\frac{d^{1+n}U_\infty^{1+n/2}}{16^n \pi^{(1-n)/2}(\nu x)^{1+n/2}}\exp(-\frac{4\gamma\rho}{JU_\infty}x$$

$$-\frac{y^2 U_\infty}{4\nu x}) - y\frac{\gamma K}{8}\frac{d^{1+n}U_\infty^{1+n/2}}{4^n \pi^{(1+n)/2}}(\nu x)^{-\frac{5+n}{2}}(3 - n - \qquad (4.1.12)$$

$$\frac{y^2 U_\infty}{2(1+n)\nu x}) + \frac{JU_\infty}{4\gamma\rho}[1 - \exp(-\frac{4\gamma\rho}{JU_\infty}x)].$$

Here $\beta = 2\eta + \gamma - \nu$. We note, that when $\beta = 0$, expressions (4.1.11) and (4.1.12) become considerably simplified. On the other hand, when $\omega = 0$, equation (4.1.3) turns into the Prandtl equation of momentum turbulent transfer and equation (4.1.4) turns into the Taylor equation of vortex turbulent transfer. The requirement of their coincidence means (compare with [278]) the realization of equality $2\eta = \nu$.

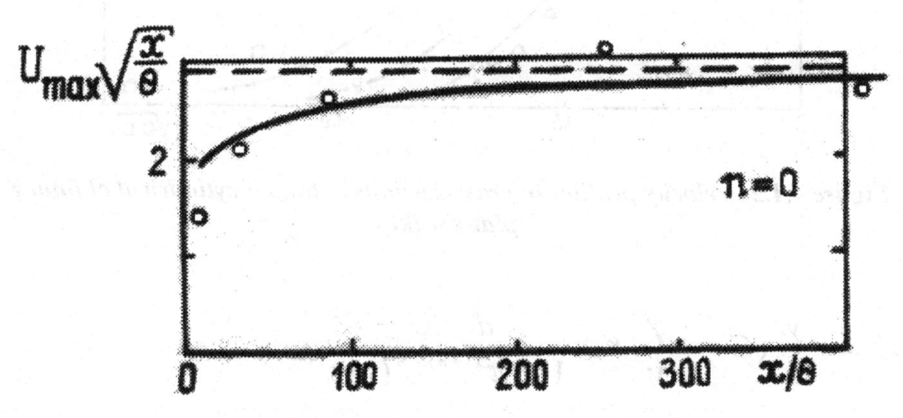

Figure 4.1.1. Maximum velocity profile in a plane wake

Therefore, we assume $\beta = 0$ in accordance with the foregoing remarks. Then the calculation formulae will get the simpler form:

$$\Omega = - y(K + \frac{\gamma}{\nu}G)\frac{d^{1+n}U^{1+\frac{n}{2}}}{2\times 16^n \pi^{(1-n)/2}(\nu x)^{1+n/2}}\times \qquad (4.1.13)$$

$$\exp\left(-\frac{U_\infty y^2}{4\nu x}\right) + y\,\frac{\gamma G\, d^{1+n}\, U_\infty^{1+\frac{n}{2}}\, (\nu\, x)^{-\frac{5+n}{2}}}{8 \times 4^n\, \pi^{(1-n)/2}} \times$$

$$\left[3 - n + \frac{U_\infty y^2}{(1+n)2\nu x}\right] - \frac{J\, U_\infty}{4\gamma}\left[1 - \exp\left(-\frac{4\gamma}{J\, U_\infty}\, x\right)\right],$$

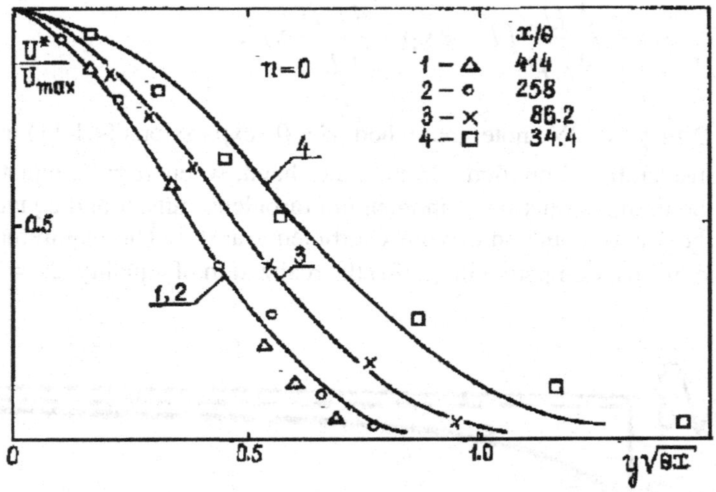

Figure 4.1.2. Velocity profiles in cross-sections behind a cylindrical chimney (plane wake)

$$\omega = \frac{y}{2}\left(G - \frac{\gamma}{\nu} K\right)\frac{d^{1+n} U_\infty^{1+n/2}}{16^n \pi^{(1-n)/2} (\nu x)^{1+n/2}}$$

$$\exp\left(-\frac{4\gamma}{JU_\infty}\, x - \frac{y^2 U_\infty}{4\nu x}\right) - \frac{y}{8}\,\frac{\gamma\, K\, d^{1+n} U_\infty^{1+n/2}}{4^n \pi^{(1+n)/2}} \times \tag{4.1.14}$$

$$(\nu x)^{-(5+n)/2}\left(3 - n - \frac{y^2 U_\infty}{2(1+n)\nu x}\right) +$$

$$\frac{JU_\infty}{4\gamma}\left\{1 - \exp\left(-\frac{4\gamma}{JU_\infty}\, x\right)\right\}$$

Integrating (4.1.13) over y, we find the profile of average velocities

$$U^{\Delta} = \frac{d^{1+n} U_{\infty}^{(1+n)/2}}{16^n \pi^{(1-n)/2} (\nu x)^{(1+n)/2}} \exp(-\frac{U_{\infty} y^2}{4\nu x})\{(K + \frac{\gamma}{\nu} G)$$

$$- \frac{1}{2^{1-n}} \frac{\gamma}{\nu} G \frac{J U_{\infty}}{4\gamma x} [1 - \exp(-\frac{4\gamma}{J U_{\infty}})][1 - \frac{y^2 U_{\infty}}{2(n+1)\nu x}]\}. \qquad (4.1.15)$$

The body resistance W will be determined by the formula

$$W = 2\rho U_{\infty} \int_0^{\infty} (\pi y)^n U^{\Delta} dy = \frac{2}{8^n} \rho \pi^n d^{1+n} (K + \frac{\gamma}{\nu} G) U_{\infty}^{1+n}. \qquad (4.1.16)$$

Thus the spin angular velocity of eddies ω leads to the body-effective resistance increase.

We note that assumption of the source Q for the field U and of the dipole for the ω in the role of boundary conditions [216] does not allow us to discern this effect.

Before comparison with experiment, let us pay attention to the fact that the integral of the nonlinear equations of the boundary layer theory, connected with a body resistance, is true inside a wake of body:

$$\frac{W}{2\rho U_{\infty}^2 \pi^n} = \int_0^{\infty} \frac{U_n}{U_{\infty}} (1 - \frac{U_n}{U_{\infty}}) y^n dy = \Sigma, \qquad (4.1.17)$$

Here U_n is a solution of nonlinear equations.

But if one calculates integral (4.1.17) with the solution U^{Δ} of linear equations, found above, then the result will be somewhat different:

$$\int_0^{\infty} \frac{U^{\Delta}}{U_{\infty}} (1 - \frac{U^{\Delta}}{U_{\infty}}) y^n dy = \Sigma_0 . \qquad (4.1.18)$$

Note that $\Sigma = \Sigma_0$ when $U_n / U_{\infty} << 1$. However, it is not so nearby the body. Therefore, we introduce a function

$$U^* = U^*(\ x;\ y\ \sqrt[1+n]{\frac{\Sigma_0}{\Sigma}}\ ). \tag{4.1.19}$$

The thickness θ of an impulse loss at the back edge of a streamlined plate will be used also:

$$\theta = (K + \frac{\gamma}{\nu} G)(\frac{d}{U_\infty}). \tag{4.1.20}$$

where d is the plate thickness. It is easy to verify that U^* satisfies equality (4.1.17), and now we shall compare this corrected solution $U^*(x, y)$ with experimental data [45, 46, 89].

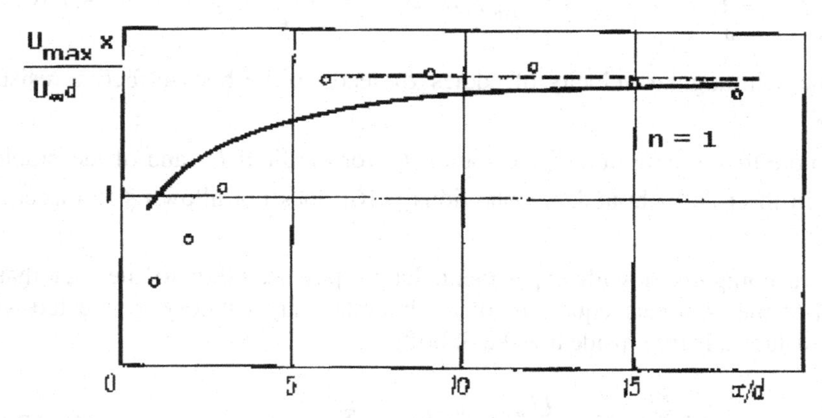

Figure 4.1.3. Maximum velocity in a wake behind a body of revolution

The profiles of the velocity distribution (Figures 4.1.1 – 4.1.4) are given here. The experimental data are marked by symbols. Solution (4.1.19) is represented as solid lines. The correcting parameters were found from data along the wake axis.

The profile of velocity maximum inside the plane wake, multiplied by $\sqrt{x}$, is represented in Figure 4.1.2. The hypothesis of similarity [31, 96] leads to the straight line, designated by the broken line.

One can see from Figures 4.1.2 and 4.1.3 that the curves reach the self - similar dependence inside the wake of the plate when $x/d = 100$.

It follows from Figures 4.1.3 and 4.1.4, that inside the wake of the body of revolution, the maximum velocity submits to the hyperbolic law and the lateral velocity distribution is well described by the Gauss curve, starting from $x/d = 6$.

Here d is a diameter of the middle cross-section of the body of revolution with the half - axis ratio equal to 1/6.

We has selected the following set of the parameter values for a plane case ($n = 0$):

$$\frac{\nu}{\theta U_\infty} = 0.038, \quad \frac{G \gamma d}{2 \nu U_\infty} = 0.38, \quad \frac{4 \gamma \rho d}{J U_\infty} = 0.027, \qquad (4.1.21)$$

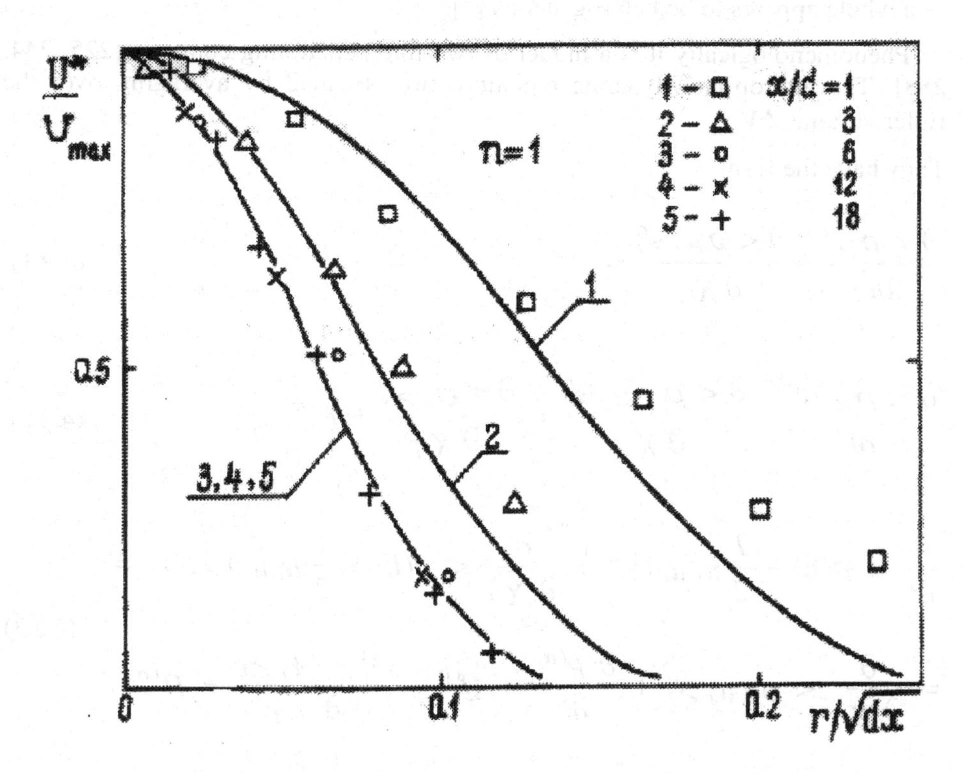

Figure 4.1.4. Velocity profiles in a cross-section of a wake behind a body of revolution

For an axisymmetric case ($n = 1$), we have

$$\frac{\gamma G}{\nu}\left(K + \frac{\gamma}{\nu} G\right)^{-1} = 0.06; \quad \frac{d}{16 \nu}\left(K + \frac{\gamma}{\nu} G\right) = 1.8$$

$$\frac{U_\infty d}{4 \nu} = 200; \quad \frac{4 \gamma d}{J U_\infty} = 0.6 \qquad (4.1.22)$$

4.2. ASYMMETRICAL DYNAMICS OF A SUSPENSION

Spatial averaging admits the simple generalization of the solution given above for the case when solid particles are suspended in a fluid, i.e., when the elementary macro volume ΔV contains two phases ($\alpha = 1,2$). Then the system as a whole appears to be heterogeneous [3].

Phenomenologically it is a model of two interpenetrating continua [225, 244, 288]. The appropriate dynamic equations are obtained by averaging over the Euler volume ΔV.

They have the form

$$\frac{\partial <\rho>^{\alpha}}{\partial t} + \frac{\partial <\rho u_j>_j^{\alpha}}{\partial X_j} = \kappa^{(\alpha)}, \tag{4.2.1}$$

$$\frac{\partial <\rho u_i>^{\alpha}}{\partial t} + \frac{\partial <\rho u_i u_j>_j^{\alpha}}{\partial X_j} = \frac{\partial <\sigma_{ij}>_j^{\alpha}}{\partial X_j} + F_i^{(\alpha)} \tag{4.2.2}$$

$$\begin{aligned} &\frac{\partial}{\partial t} < \rho(E + \frac{1}{2} u_i u_i) >^{\alpha} + \frac{\partial}{\partial X_j} < \rho(E + \frac{1}{2} u_i u_i) u_j >_j^{\alpha} = \\ &= \frac{\partial}{\partial X_j} < \sigma_{ij} u_j >_j^{\alpha} + \frac{\partial P^{(\alpha)}}{\partial t} + Q^{(\alpha)} - \frac{\partial < q_j >_j^{\alpha}}{\partial X_j} + H^{(\alpha)}. \end{aligned} \tag{4.2.3}$$

Here $\kappa^{(\alpha)}$ is the mass flux into the α-phase from coexisting phases, $F_i^{(\alpha)}$ is the bulk interphase force onto the α-phase from the co-existing phases ($\sum_\alpha F_i^{(\alpha)} = 0$), $\partial P^{(\alpha)}/\partial t$ is the work of other phases over the α-phase ($\sum_\alpha \partial P^{(\alpha)}/\partial t = 0$), $Q^{(\alpha)}$ is the internal heat source, $H^{(\alpha)}$ is the heat flux from the other phases ($\sum_\alpha H^{(\alpha)} = 0$).

Expressions (4.2.1) – (4.2.3) include the integrals over part $\Delta V^{(\alpha)}$ of the elementary volume $\Delta V = \sum_\alpha \Delta V^{(\alpha)}$, occupied by the α-phase, and correspondingly over parts $\Delta S_j^{(\alpha)}$ of the sides ΔS_j. For example,

$$< \rho\, u_i >^{\alpha} = \frac{1}{\Delta V} \int_{\Delta V} \rho\, u_i\, G^{(\alpha)}\, dV = m^{(\alpha)} \rho^{(\alpha)}\, U_i^{(\alpha)}$$

$$< \sigma_{ij} >_j^{\alpha} = \frac{1}{\Delta S_j} \int_{\Delta S_j} \sigma_{ij}\, G^{(\alpha)}\, d\, S_j = \vartheta^{(\alpha)} \sigma_{ij}^{(\alpha)}, \tag{4.2.4}$$

The integrals along the inner boundary between the phases $S_*^{(\alpha)}$ determine [225] the interphase force and work mentioned above:

$$R_i^{(\alpha)} = \frac{1}{\Delta V} \int_{\Delta S_*^{(\alpha)}} \sigma_{ij}\, d\, S_j\, ,$$

$$\frac{\partial p(\alpha)}{\partial t} = \frac{1}{\Delta V} \int_{\Delta S_*^{(\alpha)}} \sigma_{ij}\, u_i\, d\, S_j\, , \tag{4.2.5}$$

If there is no phase transition then $\kappa^{(\alpha)} = 0$ and the velocity of displacement of the boundary $S_*^{(\alpha)}$ coincides with that of the particles in the expression for the interphase work.

We used the "discriminate" function such that $G^{(\alpha)}(x_i, t) = 1$ if the particle of α phase is located at the considered micropoint and $G^{(\alpha)}(x_i, t) = 0$ if this is a particle of another phase [211, 225]. But all the phases are located in macropoint X_i, t at the same time, their presence being characterized by the surface $\vartheta^{(\alpha)}$ and volume $m^{(\alpha)}$ porosity:

$$\vartheta_j^{(\alpha)} = \frac{1}{\Delta S_j} \int_{\Delta S_j} G^{(\alpha)}\, d\, S_j = \frac{\Delta S_j^{(\alpha)}}{\Delta S_j}\, ,$$

$$m^{(\alpha)} = \frac{1}{\Delta V} \int_{\Delta V} G^{(\alpha)}\, d\, S_j = \frac{1}{\Delta X_j} \int_{-\Delta X_j/2}^{\Delta X_j/2} \vartheta_j^{(\alpha)}\, d\, x_j = \frac{\Delta V_\alpha}{\Delta V}\, . \tag{4.2.6}$$

The trustworthiness of the equality $\vartheta^{(\alpha)} = \vartheta_j^{(\alpha)} = m^{(\alpha)}$ is discussed in many publications [219, 225, 238]. This condition is used practically always, although there are methods of independent measurements of both values. Moreover, the result of spatial averaging may depend on a space dimension [70].

Let us now apply the transformation from micro-variable f to macro-variables in the case of a smooth field to find the difference between the averaging over the volume and over the surface as it follows from the integrals of (4.2.6) type.

If the surface-average value f is regular, then:

$$< f >_j = < f >_j (X_j) + \frac{\partial < f >_j}{\partial X_j}(x_j - X_j) + \frac{1}{2}\frac{\partial^2 < f >_j}{\partial X_j^2}(x_j - X_j)^2 + ... \tag{4.2.7}$$

One can see that

$$< f > = < f >_j + \frac{\partial < f >_j}{\partial X_j}\frac{1}{\Delta X_j}\int_{-\Delta X_j/2}^{\Delta X_j/2}(x - X)dx + \frac{1}{2}\frac{\partial^2 < f >_j}{\partial X_j^2}\frac{1}{\Delta X_j}\int_{-\Delta X_j/2}^{\Delta X_j/2}(x - X)^2 dx + ..., \tag{4.2.8}$$

The result of integration (4.2.8)

$$< f > = < f >_j + \frac{1}{3}\frac{\partial^2 < f >_j}{\partial X_j^2}(\Delta X_j)^2 + ... \tag{4.2.9}$$

means [221] that the difference of volume - and surface - averaged values, related to the mass-center of the volume ΔV, has the order of the square of its length scale. The limit transfer $L^{-1} \Delta X_j \rightarrow 0$ has sense only for a problem with sufficiently large scale L. Besides the condition $\Delta X_j >> \lambda$ defines the lower permissible limit for ΔX_j, where λ is a mesostructure scale (of a suspended particle or an eddy) that forbids the transfer to a micropoint.

Nevertheless, the continuum approximation is fully permissible in many cases as long as the simultaneous limit transfer from the finite differences $\Delta < f >$,... to the differentials $d < f >$,... is justified by increase of external scale L.

If the microstress σ_{ij} plays a role of f, then condition (4.2.9) means that the volume - averaged stresses $< \sigma_{ij} >$ and the macrostress $< \sigma_{ij} >_j$, in accordance with the Cauchy stress concept, differ only by the value of square order $O(\Delta X)^2$. The difference between these two average values could seem to become negligible under the limit transfer when $\Delta X_j \rightarrow 0$. However, this would be the return transition to the level below a mesostructure.

Therefore, the macrostress $< \sigma_{ij} >_j$ may be found to be nonsymmetric even under the symmetry of the microtensor σ_{ij} due to the mesostructure presence, although $< \sigma_{ij} >$ is also symmetrical. Under that symmetry, multiplying of equality (4.2.9) by the Levi-Civita alternator leads to the following result

$$\varepsilon_{kij} < \sigma_{ij} >_j \sim O(\Delta X)^2 , \tag{4.2.10}$$

which conforms the arguments of Section 2.2.

According to evaluation (4.2.10), the skew part of the macrostresses has the order of λ^2 at least [222], that is, it has the order of the specific moment of inertia J of a suspended particle (or of an eddy). Thus, angular momentum M_i is proportional to $N\lambda^2$ where N is a number of suspended particles inside the volume ΔV [3]. Although the λ scale is small, the value M_i is not negligible at all in comparison with other terms (the angular velocity may be high).

Correspondingly, the equations for the angular moments of phases are obtained [3, 219] by spatial averaging for the suspensions of rotating particles and have a form:

$$C_i^{(1)} + Z_i + \varepsilon_{ijk} < \sigma_{kj} >_j^{(1)} = 0, \tag{4.2.11}$$

$$C_i^{(2)} - Z_i + \frac{\partial \mu_{ij}^{(2)}}{\partial X_j} = 2\rho J^{(2)} \left(\frac{\partial < \Phi_i >^{(2)}}{\partial t} + U_j^{(2)} \frac{\partial < \Phi_i >^{(2)}}{\partial X_j} \right)$$

Here $2 < \Phi_i >^{(2)} = \Omega_i^{(2)} + \omega_i^{(2)}$, $C_i^{(\alpha)}$ are the external volume-distributed moments and created, for example, by the magnetic field [264, 272, 290]; phase *(1)* corresponds to a suspending fluid and phase *(2)* – to suspended particles.. The couple interaction Z_i, between the phases, is evaluated by the solution for the solid spherical particle with a radius λ, rotating with angular velocity ω_i relatively to ambient fluid ($\alpha = 1$) with viscosity ν:

$$Z_i = \gamma\rho\,\omega_2, \tag{4.2.12}$$

$$\gamma = 6\nu(1 - m) = 8\pi\, r^3\, \nu N,$$

Here N is the number of suspended particles inside the volume unit; moment of inertia $J = (2/5)\, r^2 \sim r^2 / 2 =$ corresponds to unit volume of the sphere and the rotation viscosity were determined in [3]. The latter is linearly proportional to the volume concentration $m^{(2)} = 1 - m$ of solid phase, $m^{(1)} = m$.

Let us mention the fact that, when $C_i^{(1)} + C_i^{(2)} = C_i$, $U_i^{(1)} = U_i^{(2)}$, system (4.2.11), (4.2.12) can be represented in the form of the angular momentum balance discussed in Chapter 2:

$$\rho\, J^{(2)} [\frac{\partial}{\partial t}(\Omega_i^{(2)} + \omega_i^{(2)}) + U_j \frac{\partial}{\partial X_j}(\Omega_i^{(2)} + \omega_i^{(2)})] =$$

$$\varepsilon_{ijk} < \sigma_{ij} >_j^{(1)} + C_i + \frac{\partial \mu_{ij}^{(2)}}{\partial X_j} \tag{4.2.13}$$

Expression (4.2.12) corresponds to the conventional constitutive law for the asymmetric part of a stress tensor:

$$\varepsilon_{ijk} < \sigma_{kj} >_j^{(1)} = 4\rho\gamma\,\omega_i\,. \tag{4.2.14}$$

That is why the micropolar fluid model (or, in other words, asymmetric hydrodynamics) was used in the hydrodynamics of suspensions [6, 132-135].

As to mass and impulse balances (4.2.1), (4.2.2), the following forms were suggested, see [116, 219]:

$$\frac{\partial m\,\rho_1}{\partial t} + \frac{\partial m\,\rho_1 U_j^{(1)}}{\partial X_j} = 0, \tag{4.2.15}$$

$$\frac{\partial}{\partial t}[(1-m)\rho_2]+\frac{\partial}{\partial X_i}[(1-m)\rho_2 U_i^{(2)}]=0, \tag{4.2.16}$$

$$m\,\rho_1\frac{d_1 U_1^{(1)}}{\partial t} = -m\frac{\partial p}{\partial X_i}+\frac{\partial}{\partial X_j}(\sigma_{ij}^s+\sigma_{ij}^a)-B_i\,, \tag{4.2.17}$$

$$(1-m)\,\rho_2\frac{\partial_2 U_i^{(2)}}{\partial t} = -(1-m)\frac{\partial p}{\partial X_i}+B_i\,, \tag{4.2.18}$$

Here σ_{ij}^s is the symmetrical part of the viscous stress tensor, σ_{ij}^a is its skew part and the symbol of phase substation phase derivative is used:

$$\sigma_{ij}^s = -\frac{2}{3}\rho\nu_*\frac{\partial U_i^{(1)}}{\partial X_i}+\rho\nu_*(\frac{\partial U_i^{(1)}}{\partial X_j}+\frac{\partial U_j^{(1)}}{\partial X_i}),$$

$$\frac{d_\alpha}{dt} = \frac{\partial}{\partial t}+U_j^\alpha\frac{\partial}{\partial X_j} \tag{4.2.19}$$

The constitutive laws include the Einstein definition [73] of disturbed fluid viscosity:

$$\nu_* = \nu_0\,[1+\frac{5}{2}(1-m)] \tag{4.2.20}$$

The following interphase bulk force, compare [219, 225], was used in (4.2.17), (4.2.18):

$$F_i \equiv F_i^{(1)} = -F_i^{(2)} = -p\frac{\partial m}{\partial X_i}+B_i$$

It includes the viscous and the lift bulk forces:

$$B_i = a(U_i^{(1)}-U_i^{(2)})-b\,\varepsilon_{kij}(U_j^{(1)}-U_j^{(2)})(\Phi_k^{(2)}-\Omega_k^{(1)}),$$

$$a = \frac{9}{2\lambda^2}\rho\nu_0 m(1-m),\qquad b=\frac{3}{4}\rho_1(1-m),$$

The Einstein viscosity is valid for the values m, close to one, (for dilute suspension [73]) and has to be modified for other values of m.

The viscous part B_i of interaction force F_i between the phases can also include the Saffman lifting force [255]. Equations (4.2.11), (4.2.18) can include the stresses created by solid particle interaction (see [219, 225] for example).

The applications of the model were discussed (see [209, 219], for example) and they are not repeated here.

It was suggested in [209] that introduction of skew stress σ_{ij}^a be avoided (4.2.14), replacing its divergence by the equivalent vector

$$A_i = \frac{\partial \sigma_{ij}^a}{\partial X_j} \tag{4.2.21}$$

Then A_i can be included into the interaction force between phases ($B_i^* = A_i + B_i$). This means actually another determination of the latter.

However, A_i and B_i have different origin: the first one reflects action of the outer (for ΔV) continuum but the second one is internal for the system included in ΔV.

That is why B_i does not appear in the momentum balance for the whole system (in contrast to A_i, that is, to surface force, corresponding to σ_{ij}^a). But if one believes that just B_i^* is the internal force, then σ_{ij}^a actually disappears from the total impulse balance, although the corresponding models of suspension dynamics [209] will turn out to be more limited than as it is formulated here. They will correspond to the internal mutual balance of bulk moments distributed over the phases.

The method of introducing the volume force A_i instead of stress divergence σ_{ij} is well known itself (see [153], for example).

It was noted [220, 222] that an incorrect conclusion about the necessary symmetry of the elastic stress tensor was obtained in [153]. The error happened to be in the statement that the external couple forces action, which follows from the angular momentum equilibrium (compare with the Cosserat equation) for any finite body:

$$M_{ik} = \int_S (\sigma_{il} x_k - \sigma_{kl} x_i) d S_l + \int_V (\sigma_{ki} - \sigma_{ik}) dV, \tag{4.2.22}$$

can be reduced to the surface integral only in the case when

$$\sigma_{ik} = \sigma_{ki} \tag{4.2.23}$$

G. Batchelor [22] shared this point of view to find a stress tensor in a suspension. But (4.2.23) is the simplest form of the angular momentum balance (2.3.9) and it is quite possible that

$$\sigma_{ki} - \sigma_{ik} = \frac{\partial \mu_{ikj}}{\partial X_j} \tag{4.2.24}$$

Therefore, the asymmetrical component of the stress tensor can be expressed by the divergence of a higher rank tensor (of the Mindlin double stresses [187]). Then the required condition of only surface action will be realized even at $\sigma_{ki} \neq \sigma_{ik}$ [16].

In the last edition of the corresponding part of the cited monograph [17], equality (4.2.23) was renamed as a particular case. However, it is worth noting the discussions [12, 109] concerning the Landau theory of two components in liquid helium without using angular momentum balance in a combination with the Feinman idea of quantum vortex lines. Experimentally (E. Andronikoshvili et al, see [242]) it was shown that the molecular angular rotation exists and this effect is described as quantum circulation [242, 280] and is indeed needed in the angular momentum balance indeed [12, 109].

4.3. TURBULENCE OF HETEROGENEOUS FLUIDS

Let us consider now the equations of dynamics of turbulized heterogeneous fluid [116]. The difference consists in the fact that not only the rotating solid particles but also the fluid eddies (the rotating Prandtl's moles) is present in a suspension due to its turbulization.

The effect of asymmetry of the stress tensor is incomparably more essential whereas the viscous stresses can be neglected if heavy particles are suspended in a flow. Many actual problems of atmospheric turbulence are connected with dispersion of particles or liquid (rain) drops in air.

In accordance with the procedure of spatial averaging, we have:

$$\frac{\partial}{\partial t} < m\, \rho_1 > + \frac{\partial}{\partial X_j} < m\, \rho_1\, u_j^{(1)} >_j \; = 0 \tag{4.3.1}$$

[16] I. Tamm insisted on Maxwell stress asymmetry in his "Theory of Electricity" [274].
[17] L.D. Landau, E. M. Lifshitz. "Theory of elasticity." The 4th edition. Moscow: Nauka, 1987, p. 17 [154].

$$\frac{\partial}{\partial t} < (1 - m)\,\rho_2 > + \frac{\partial}{\partial X_j} < (1 - m)\,\rho_2\, u_j^{(2)} >_j = 0$$

$$\frac{\partial}{\partial t} < m\,\rho_1\, u_i^{(1)} > + \frac{\partial}{\partial X_j} < m\,\rho_1\, u_i^{(1)}\, u_j^{(1)} >_j = -m\frac{\partial p}{\partial X_i} - < B_i >$$

(4.3.2)

$$\frac{\partial}{\partial t} < (1 - m)\,\rho_2\, u_i^{(2)} > + \frac{\partial}{\partial X_j} < (1 - m)\,\rho_2\, u_i^{(2)}\, u_j^{(2)} >_j =$$

$$= -(1 - m)\frac{\partial p}{\partial X_i} + < B_i >$$

$$\frac{\partial}{\partial t} < \varepsilon_{ijk}\, x_k\, m\,\rho_1\, u_j^{(1)} > + \frac{\partial}{\partial X_j} < \varepsilon_{ilk}\, x_k\, m\,\rho_1\, u_l^{(1)}\, u_j^{(1)} >_j =$$

$$= -m\frac{\partial}{\partial X_j} < \varepsilon_{ilk}\, x_k\, p\,\delta_{lj} >_j - < \varepsilon_{ijk}\, x_k\, B_j >,$$

(4.3.3)

$$\frac{\partial}{\partial t} < \varepsilon_{ijk}\, x_k\, (1 - m)\rho_2\, u_j^{(2)} > + \frac{\partial}{\partial X_j} < \varepsilon_{ilk}\, x_k\, (1 - m)\rho_2\, u_l^{(2)} u_j^{(2)} >_j =$$

$$= -(1 - m)\frac{\partial}{\partial X_j} < \varepsilon_{ilk}\, x_k\, p\,\delta_{lj} >_j + < \varepsilon_{ijk}\, x_k\, B_i > .$$

Now we represent the instant velocity fields as

$$u_i^{(\alpha)}(\xi_j, t) = U_i^{(\alpha)} + \frac{\partial U_i^{(\alpha)}}{\partial X_j}\xi_j + O\left(\frac{\Delta^2}{L^2}\right) + w_i^{(\alpha)}(\xi_j, t), \qquad (4.3.4)$$

Here $U_i^{(\alpha)}$ is the velocity of the phase α, related to the mass center of the volume ΔV, $w_i^{(\alpha)}$ is a turbulent pulsation of the phase α. Let us divide the volume ΔV again into a number of subvolumes ΔV_α. The representation

$$w_j^{(\alpha)}(\xi_m, t) = w_j^{(\alpha)} + (\partial w_j^{(\alpha)} / \partial \overline{\xi}_m^{(\alpha)})\zeta_m^{(\alpha)},$$

$$\xi_m = \overline{\xi}_m^{(\alpha)} + \zeta_m^{(\alpha)},$$

$$w_j^{(\alpha)} = w_j(\overline{\xi}_m^{(\alpha)}) \tag{4.3.5}$$

is valid in ΔV_α. Here the coordinates $\varsigma_m^{(\alpha)}$ relative to the mass center of the mole volume ΔV_α of phase α are used.

Let us introduce the "geometrical" inertia moment of phase α:

$$i_{mn}^{(\alpha)} = \frac{1}{\Delta V_\alpha} \int_{\Delta V_\alpha} \zeta_m \zeta_n \, dV \tag{4.3.6}$$

If the flow is turbulent with sufficiently high spin velocities, we have the following expressions for the angular momentum:

$$M_i^{(1)} = < \varepsilon_{ilk} \rho u_l \xi_k >^{(1)} = m \rho_1 < \varepsilon_{iln} (\frac{\partial U_l^{(1)}}{\partial X_m} + \frac{\partial w_l^{(1)}}{\partial \overline{\xi}_m}) i_{mn}^{(1)} >,$$

$$M_i^{(2)} = < \varepsilon_{ilk} \rho u_l \xi_k >^{(2)} = (1 - m) \rho_2 < \varepsilon_{iln} (\frac{\partial U_l^{(2)}}{\partial X_m} + \frac{\partial w_l^{(2)}}{\partial \overline{\xi}_m}) i_{mn}^{(2)} >, \tag{4.3.7}$$

Accordingly, the expressions for kinetic moment fluxes are valid:

$$< \varepsilon_{ilk} \rho u_l \xi_k u_j >_j^{(1)} = M_i^{(1)} U_j^{(1)} - \mu_{ij}^{(1)} + O(\frac{\Delta}{L}),$$

$$< \varepsilon_{ilk} \rho u_l \xi_k u_j >_j^{(2)} = M_i^{(2)} U_j^{(2)} - \mu_{ij}^{(2)} + O(\frac{\Delta}{L}). \tag{4.3.8}$$

If rotating particles are symmetrical, then

$$i_{mn}^{(\alpha)} = \frac{1}{2} i^{(\alpha)} \delta_{mn}$$

Let $\omega_i^{(\alpha)}$ be the spin. Then

$$M_i^{(1)} = J^{(1)} m \rho_1 (\Omega_i^{(1)} + \omega_i^{(1)}),$$
$$M_i^{(2)} = J^{(2)} (1 - m) \rho_2 (\Omega_i^{(2)} + \omega_i^{(2)}) \qquad (4.3.9)$$

Here $J^{(\alpha)} = < i^{(\alpha)} >$ is an average inertia moment of the phase α. For simplicity, assume that the specific inertia moments of moles of both phases are equal $J^{(1)} = J^{(2)} = J$. It corresponds to the same linear scale of suspended particles and of turbulent eddies[18].

Now the equations of dynamics of turbulent suspension can be represented as:

$$\frac{\partial}{\partial t}(m \rho_1) + \frac{\partial}{\partial X_j}(m \rho_1 U_j^{(1)}) = 0,$$
$$\frac{\partial}{\partial t}(1 - m) \rho_2 + \frac{\partial}{\partial X_j}(1 - m) \rho_2 U_j^{(2)} = 0; \qquad (4.3.10)$$

$$m \rho_1 \frac{d_1 U_i^{(1)}}{d t} = \frac{\partial}{\partial X_j} m \rho_1 R_{ij}^{(1)} - m \frac{\partial p}{\partial X_i} - am(U_i^{(1)} - U_i^{(2)}),$$
$$(1 - m) \rho_2 \frac{d_2 U_i^{(2)}}{d t} = \frac{\partial}{\partial X_j}(1 - m) \rho_2 R_{ij}^{(2)} - (1 - m) \frac{\partial p}{\partial X_i} + \qquad (4.3.11)$$
$$+ am(U_i^{(1)} - U_i^{(2)});$$

$$\frac{\partial}{\partial t} M_k^{(1)} + \frac{\partial}{\partial X_l}(M_k^{(1)} U_l^{(1)} - \mu_{kl}^{(1)}) = \varepsilon_{kij} R_{ij}^{(1)} m +$$
$$+ am[(\Omega_k^{(2)} + \omega_k^{(2)}) - (\Omega_k^{(1)} + \omega_k(1))], \qquad (4.3.12)$$
$$\frac{\partial}{\partial t} M_k^{(2)} + \frac{\partial}{\partial X_l}(M_k^{(2)} U_l^{(2)} - \mu_{kl}^{(2)}) = \varepsilon_{kij} R_{ij}^{(2)} (1 - m) -$$
$$- am[(\Omega_k^{(2)} + \omega_k^{(2)}) - (\Omega_k^{(1)} + \omega_k^{(1)})].$$

[18] Other variants of particle/eddy scales were estimated in [59, 60, 142].

Here $R_{ij}^{(\alpha)}$ are the Reynolds phase stresses, $\Omega_i^{(\alpha)}$ is the eddy of the field of phase velocity; the effect of lifting force was not taken into consideration (coefficient $b = 0$). We shall introduce the closing relations for system (4.3.10) as follows:

$$\frac{1}{2}(R_{ij}^{(\alpha)} + R_{ji}^{(\alpha)}) = \nu^{(\alpha)} \rho_\alpha (\frac{\partial U_i^{(\alpha)}}{\partial X_j} + \frac{\partial U_j^{(\alpha)}}{\partial X_i}),$$

$$\frac{1}{2}(R_{ij}^{(\alpha)} - R_{ji}^{(\alpha)}) = \varepsilon_{ijn}\, 2\, \gamma^{(\alpha)} \rho_\alpha\, \omega_n^{(\alpha)}, \tag{4.3.13}$$

$$\mu_{ij}^{(\alpha)} = 2\, \eta^{(\alpha)} \rho_\alpha\, J^{(\alpha)} \frac{\partial}{\partial X_j}(\Omega_i^{(\alpha)} + \omega_i^{(\alpha)}),$$

Here $\nu^{(\alpha)}, \eta^{(\alpha)}, \gamma^{(\alpha)}$ are the coefficients of turbulent phase viscosities.

We linearize the equations around the infinity state and below all variable values are deviations from that state.

Let us consider a turbulent cross-section of suspension in the plane wake of the body. That is, we represent the velocity components and the volume fluid concentration in the form:

$$U_1^{(\alpha)} = U_\infty - \overline{U}^{(\alpha)}, \ m = m_\infty - \overline{m}, \ X_1 = x; \ X_2 = y.$$

where bar means a total value. We shall also denote

$$\omega^{(\alpha)} = \omega_3^{(\alpha)}, \quad \Omega^{(\alpha)} = \Omega_3^{(\alpha)}.$$

In the Oseen approximation system (4.3.10), (4.3.11) should be written as

$$U_\infty \frac{\partial U^{(1)}}{\partial x} - \nu^{(1)} \frac{\partial^2 U^{(1)}}{\partial y^2} = -2\, \gamma^{(1)} \frac{\partial \omega^{(1)}}{\partial y} + \frac{a(m_\infty)}{\rho_1}(U^{(2)} - U^{(1)})$$

$$U_\infty \frac{\partial \omega^{(1)}}{\partial x} - 2\, \eta^{(1)} \frac{\partial^2 \omega^{(1)}}{\partial y^2} + 4 \frac{\gamma^{(1)}}{J} \omega^{(1)} = \tag{4.3.14}$$

$$= (2\eta^{(1)} - \nu^{(1)})\frac{\partial^2 \Omega^{(1)}}{\partial y_2} + \frac{a(m_\infty)}{\rho_1}(\omega^{(2)} - \omega^{(1)}),$$

$$U_\infty \frac{\partial U^{(2)}}{\partial x} - \nu^{(2)}\frac{\partial^2 U^{(2)}}{\partial y^2} = -2\gamma^{(2)}\frac{\partial \omega^{(2)}}{\partial y} +$$

$$\frac{a(m_\infty)\, m_\infty}{\rho_2 (1 - m_\infty)}(U^{(1)} - U^{(2)}),$$

$$U_\infty \frac{\partial \omega^{(2)}}{\partial x} - 2\eta^{(2)}\frac{\partial^2 \omega^{(2)}}{\partial y^2} + 4\frac{\gamma^{(2)}}{J}\omega^{(2)} =$$

$$= (2\eta^{(2)} - \nu^{(2)})\frac{\partial^2 \Omega^{(2)}}{\partial y^2} + \frac{a(m_\infty)\, m_\infty}{\rho_2 (1 - m_\infty)}(\omega^{(1)} - \omega^{(2)}).$$

The streamlined body with characteristic size of d can be modeled by dipoles. The corresponding boundary conditions are formulated again in the following way:

$$\Omega^{(\alpha)} = K^{(\alpha)}[\delta(y + \frac{d}{2}) - \delta(y - \frac{d}{2})],$$

$$\omega^{(\alpha)} = -G^{(\alpha)}[\delta(y + \frac{d}{2}) - \delta(y - \frac{d}{2})],$$

At infinity the rest conditions are valid:

$$\left.\begin{aligned} U^{(\alpha)} &= \frac{\partial U^{(\alpha)}}{\partial y} = 0, \\ \omega^{(\alpha)} &= \frac{\partial \omega^{(\alpha)}}{\partial y} = 0. \end{aligned}\right\} \quad |y| \to \infty \tag{4.3.15}$$

Assume that the asymmetrical stress components are much smaller than the symmetrical ones, i.e. $|\nu^{(\alpha)}\Omega^{(\alpha)}| >> |\gamma^{(\alpha)}\omega^{(\alpha)}|$. Besides let the inequality

$$4\sqrt{\nu^{(\alpha)} x / U_\infty} >> d$$

and the equality $2\eta^{(\alpha)} + \gamma^{(\alpha)} - \nu^{(\alpha)} = 0$ is satisfied.

The inequality $\rho_1 m >> \rho_2 (1-m)$ takes place in the case of diluted suspension, and the interphase force in the impulse balance for fluid is relatively small and can be neglected.

The solution of system (4.3.11) is constructed by the method of consequential approximations, the solution for the k step being used for the approximation of the right-hand sides of system (4.3.11) at the $(k+1)$ step. Under the analogous simplifications, the equations of the first approximations have a form

$$U_\infty \frac{\partial \Omega_0^{(1)}}{\partial x} - \nu^{(1)} \frac{\partial^2 \Omega_0^{(1)}}{\partial y^2} = 0,$$

$$U_\infty \frac{\partial \omega_0^{(1)}}{\partial x} - 2\eta^{(1)} \frac{\partial^2 \omega_0^{(1)}}{\partial y^2} + \frac{4\gamma^{(1)}}{J} \omega_0^{(1)} = 0,$$

$$U_\infty \frac{\partial \Omega_1^{(1)}}{\partial x} - \nu^{(1)} \frac{\partial^2 \Omega_1^{(1)}}{\partial y^2} = -\gamma^{(1)} \frac{\partial^2 \omega_0^{(1)}}{\partial y^2} + \frac{a(m_\infty)}{\rho_1} (\Omega_0^{(2)} - \Omega_0^{(1)}),$$

$$U_\infty \frac{\partial \omega_1^{(1)}}{\partial x} - 2\eta^{(1)} \frac{\partial^2 \omega_1^{(1)}}{\partial y^2} + \frac{4\gamma^{(1)}}{J} \omega_1^{(1)} =$$

$$= (2\eta^{(1)} - \nu^{(1)}) \frac{\partial^2 \Omega_0^{(1)}}{\partial y^2} + \frac{a(m_\infty)}{\rho_1} (\omega_0^{(2)} - \omega_0^{(1)})$$

$$U_\infty \frac{\partial \Omega_0^{(2)}}{\partial x} - \nu^{(2)} \frac{\partial^2 \Omega_0^{(2)}}{\partial y^2} = \frac{a(m_\infty) m_\infty}{\rho_2 (1 - m_\infty)} (\Omega_0^{(1)} - \Omega_0^{(2)}),$$

$$U_\infty \frac{\partial \omega_0^{(2)}}{\partial x} - 2\eta^{(2)} \frac{\partial^2 \omega_0^{(2)}}{\partial y^2} + \frac{4\gamma^{(2)}}{J} \omega_0^{(2)} \frac{a(m_\infty) m_\infty}{\rho_2 (1 - m_\infty)} (\omega_0^{(1)} - \omega_0^{(2)}),$$

(4.3.16)

$$U_\infty \frac{\partial \Omega_1^{(2)}}{\partial x} - \nu^{(2)} \frac{\partial^2 \Omega_1^{(2)}}{\partial y^2} = \frac{a(m_\infty)\, m_\infty}{\rho_2 (1 - m_\infty)} (\Omega_0^{(1)} - \Omega_0^{(2)}) - \gamma^{(2)} \frac{\partial^2 \omega_0^{(2)}}{\partial y^2},$$

$$U_\infty \frac{\partial^2 \omega_1^{(2)}}{\partial x} - 2\, \eta^{(2)} \frac{\partial^2 \omega_1^{(2)}}{\partial y^2} + \frac{4\, \gamma^{(2)}}{J} \omega_1^{(2)} =$$

$$= (2\, \eta^{(2)} - \nu^{(2)}) \frac{\partial^2 \Omega_0^{(2)}}{\partial y^2} + \frac{a(m_\infty)\, m_\infty}{\rho_2 (1 - m_\infty)} (\omega_0^{(1)} - \omega_0^{(2)});$$

Actually, the boundary conditions (4.3.12) are satisfied at each step. We accept [116] the following assumptions about the dipoles intensity. At first, the impulse of a streamlined body is passed directly to the fluid phase. Secondly, the solid particles have a higher inertia than the fluid ones.

Therefore we assume $K^{(1)} >> K^{(2)}$ and $G^{(1)} >> G^{(2)}$. Then we obtain the following expression for velocity field of the fluid phase from (4.3.13):

$$U_1^{(1)} = \frac{d\sqrt{U_\infty}}{\sqrt{\pi\, \nu^{(1)} x}} \exp\left(- \frac{U_\infty y^2}{4\, \nu^{(1)} x}\right)\{(K^{(1)} + \frac{\gamma^{(1)}}{\nu^{(1)}} G^{(1)})\} -$$

$$\frac{1}{2} \frac{\gamma^{(1)}}{\nu^{(1)}} G^{(1)} \frac{J\, U_\infty}{4\, \gamma^{(1)} x} [1 - \exp(- \frac{4\, \gamma^{(1)} x}{J\, U_\infty})](1 - \frac{U_\infty y^2}{2\, \nu^{(1)} x}) +$$

$$+ \frac{a(m_\infty)}{\rho_1 U_\infty} \frac{K^{(2)} d\sqrt{U_\infty}}{\sqrt{\pi}} \int_0^x \exp\{ - \frac{U_\infty y^2}{4\, \nu^{(1)}(x - t) + 4\, \nu^{(2)} t} - \tag{4.3.17}$$

$$\frac{a(m_\infty)\, m_\infty t}{\rho_2 U_\infty (1 - m_\infty)}\} \frac{dt}{\nu^{(1)}(x - t) + \nu^{(2)} t},$$

The solution for velocity of solid phase $U_1^{(2)}$ is totally analogous. If one expands the integrands of (4.3.15) and of its analogue in a series of parameter $(\nu^{(1)} - \nu^{(2)})(x - t)/(\nu^{(1)} x)$, the following estimation will be obtained:

$$U_1^{(2)} = U_1^{(1)} [1 + O(\frac{1}{x})] \exp\{- \frac{a(m_\infty) m_\infty x}{\rho_2 U_\infty (1 - m)}\}. \tag{4.3.18}$$

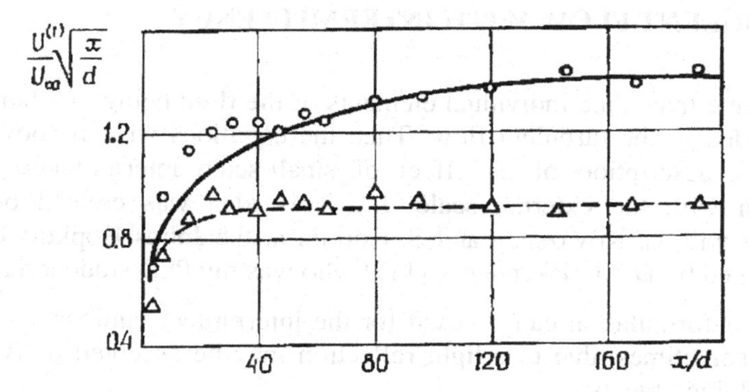

Figure 4.3.1. The axial velocity of fluid phase at the selected set of parameters in a wake behind a body

Therefore, the phase velocities become equal in the zone far from the streamlined body. However, a defect of the solid phase velocity $U_1^{(2)}$, close to the body, is bigger several times than the correspondent value $U^{(1)}$ for fluid phase.

According to the experiment [86], this difference is preserved at the comparatively far distances. It can be shown that at a great distance x the axial velocity of fluid phase decreases as $x^{-1/2}$. The deviations from this law inside the zone close to the body are connected with the fluid mesovortexes and have to become picked up by the rotation of suspended particles. Indeed, according to experimental data [86], the mentioned deviations grow with an increase of the bulk content of solid particles. Formula (4.3.15) corresponds to this experimental fact.

The profile of the axial velocity of the fluid phase is shown in Figure 4.3.1 for the following set parameters (triangles correspond to the experiment without suspended particles, circles – with particles):

$$\frac{\nu^{(1)}}{U_\infty d} = 0.0481; \quad \frac{\nu^{(2)}}{\nu^{(1)}} = 0.589; \quad \frac{\rho_2(1 - m_\infty)}{\rho_1 m_\infty} = 0.0155;$$
$$\frac{K^{(2)}}{U_\infty} = 14.2; \quad \frac{\gamma^{(1)}}{\nu^{(1)}}\frac{G^{(1)}}{U_\infty} = 0.326; \quad \frac{K^{(1)}}{U_\infty} = 0.029. \tag{4.3.19}$$

The assumed conditions correspond to the calculation represented in Figure 4.3.1 by the solid line. The experimental data are also represented there. The dotted line corresponds to the case when there are no suspended solid particles ($m = 1$).

4.4. TURBULENT FLOW WITH INTERMITTENCY

Let us assume that some individual elements of the fluid being in a laminar state are suspended in the turbulent flow. Then the equations written above become suitable for description of an effect of small-scale intermittency (small in comparison with the external scale L). This idea was created because of discussions with L. Kovazsnay and S. Corrsin at the Johns Hopkins University and developed by D. Sh. Iskenderov [117] who was my PhD student that time.

A similar formulation can be used for the intermittent laminar and turbulent spots that sometimes, due to a light reflection may be accepted individually as unknown flying objects.

The phenomenon of intermittency, typical for turbulence in general, consists in interchange of turbulent and quasi-laminar conditions inside the flow [110, 144, 145]. Either the turbulent spots, flowing in turbulent fluid, or the laminar inclusions, turning out to be «suspended», can represent this picture. Their "concentration" can be identified with the so-called coefficient of intermittency Γ. The latter is determined experimentally as turbulent pulsations time at a microprint dv divided by total time of the observation [53].

The value $\Gamma = 1$ corresponds, as one can see, to the completely developed turbulent stream.

However, there is a possibility to consider the same coefficient Γ [66, 117] as a volume concentration of laminar particles inside the flowing stream. So long as the mean velocity of a turbulent spot (of a mole), in the general case differs from the velocity of surrounding non-turbulent fluid [66, 110, 145], this justifies using the model of interpenetrating continua. In general, the phenomenon of intermittency corresponds to the case when the generation of eddies is deficient for complete turbulization of a whole flow.

Thus, spatial averaging gives the following balance equations for mass and impulse of both the "phases", of turbulent (1) and of laminar (2):

$$\frac{\partial}{\partial t} < \rho >^{(1)} + \frac{\partial < \rho u_j >_j^{(1)}}{\partial X_j} = \kappa;$$

$$\frac{\partial}{\partial t} < \rho >^{(2)} + \frac{\partial < \rho u_j >_j^{(2)}}{\partial X_j} = -\kappa;$$

(4.4.1)

$$\frac{\partial}{\partial t} < \rho u_i >^{(1)} + \frac{\partial < \rho u_i u_j >_j^{(1)}}{X_j} = -\Gamma \frac{\partial p}{\partial X_i} + \frac{\partial < \sigma_{ij} >^{(1)}}{\partial X_j} - B_i ,$$

$$\frac{\partial}{\partial t} < \rho u_i >^{(2)} + \frac{\partial < \rho u_i u_j >_j^{(2)}}{\partial X_j} = -(1 - \Gamma)\frac{\partial p}{\partial X_i} + \frac{\partial < \sigma_{ij} >^{(2)}}{\partial X_j} + B_i .$$

This set of equations is typical for interpenetrating continua. Here the source κ (mass growth rate of turbulent spots per unit of volume that could be negative to account for turbulence dissipation) is introduced as an essential term to model the phase transfer intensity.

Let suppose that the phase velocities can be represented as

$$u_i^{(1)}(\xi_j , t) = U_i^{(1)} + O(\frac{\Delta}{L}) + w_i(\xi_j , t),$$

$$u_i^{(2)}(\xi_j , t) = U_i^{(2)} + O(\frac{\Delta}{L}), \qquad (4.4.2)$$

The latter expression means that the pulsation velocities are absent inside the laminar inclusions. Then equations (4.4.1) will assume a form

$$\frac{\partial \Gamma}{\partial t} + \frac{\partial \Gamma U_j^{(1)}}{\partial X_j} = \frac{\kappa}{\rho},$$

$$\frac{\partial \Gamma}{\partial t} - \frac{\partial (1 - \Gamma) U_j^{(2)}}{\partial X_j} = \frac{\kappa}{\rho}, \qquad (4.4.3)$$

$$\rho \frac{\partial}{\partial t}(\Gamma U_i^{(1)}) + \rho \frac{\partial}{\partial X_j}(\Gamma U_i^{(1)} U_j^{(1)}) =$$

$$- \Gamma \frac{\partial p}{\partial X_i} + \frac{\partial}{\partial X_j}(\Gamma R_{ij}) - B_i ,$$

$$\rho \frac{\partial}{\partial t}(1 - \Gamma) U_i^{(2)} + \rho \frac{\partial}{\partial X_j}(1 - \Gamma) U_i^{(2)} U_j^{(2)} =$$

$$- (1 - \Gamma) \frac{\partial p}{\partial X_i} + \frac{\partial}{\partial X_j}(1 - \Gamma) \sigma_{ij}^{(2)} + B_i \,.$$

The viscous stresses $\sigma_{ij}^{(1)}$ inside the turbulent spots are much smaller than the turbulent stress $R_{ij} = - < \rho w_i w_j >_j$ and they can be neglected. We also neglect here the spin effect to avoid a complicated analysis.

Let us use the diffusion approximation in the present case in conformity with the rule [225], according to which the system moves with one volume mean velocity:

$$U_i = \Gamma U_i^{(1)} + (1 - \Gamma) U_i^{(2)}.$$

Then relative velocities of "phases" (mass flux rates) will be determined by the gradient formulas [59]:

$$U_j^{(1)} - U_j = - \frac{\nu}{A} \frac{1}{\Gamma} \frac{\partial \Gamma}{\partial X_j},$$

$$U_j^{(2)} - U_j = \frac{\nu}{A} \frac{1}{1 - \Gamma} \frac{\partial \Gamma}{\partial X_j}, \qquad (4.4.4)$$

Here A is an analogue of the Schmidt number.

The experimental and calculation data are given in Figure 4.4.1. Difference $(U_2^{(1)} - U_2)/U_\infty$ is represented by circles, $(U_2^{(2)} - U_2)/U_\infty$ is represented by squares. The coefficient of intermittency is represented by triangles. Here U_∞ is a velocity of external flow.

The coefficient of intermittency in such a stream can be represented [117] as

$$\Gamma = \frac{1}{2} [1 - \frac{2}{\sqrt{\pi}} \int_0^Y \exp(- \zeta^2) d\zeta],$$

$$Y = \frac{y - y_0}{\sigma \sqrt{2}} \qquad (4.4.5)$$

In Figure 4.4.1 solid curves are given for

$$y = X_2, \quad y_0 = 8.5\ cm, \quad \sigma = 1.5\ cm,$$

$$\nu = 0.0012\ AU_\infty \delta, \quad \delta = 10\ cm$$

The broken curves correspond to $\nu = 0.001\ A\,U_\infty\,\delta$ and the broken-dotted curves to

$$\nu = 0.0014\ A\,U_\infty\,\delta.$$

The diffusion model equations for turbulence with intermittency correspondingly turn out to be:

$$\frac{\partial U_i}{\partial X_i} = 0,$$

$$\frac{\partial \Gamma}{\partial t} + U_j \frac{\partial \Gamma}{\partial X_j} = \frac{\kappa}{\rho} + \frac{\partial}{\partial X_j}\left(\frac{\nu}{A}\frac{\partial \Gamma}{\partial X_j}\right), \qquad (4.4.6)$$

$$\rho \frac{\partial U_i}{\partial t} + \rho \frac{\partial U_i U_j}{\partial X_j} =$$

$$-\frac{\partial p}{\partial X_i} + \frac{\partial}{\partial X_j}\left[\Gamma R_{ij} + (1 - \Gamma)\,\sigma_{ij}^{(2)}\right].$$

Here the momentum flow, caused by the deviations of phase velocities from the volume average velocity, was not accounted for.

Naturally, if $\Gamma = 1$, the last momentum equation turns into the Reynolds equation, but if $\Gamma = 0$, it turns into the Navier-Stokes' one. The usual rheological laws are used here:

$$R_{ij} = \rho\nu\left(\frac{\partial U_i}{\partial X_j} + \frac{\partial U_j}{\partial X_i}\right),$$

$$\sigma_{ij}^{(2)} = \rho\,\nu_0\left(\frac{\partial U_i}{\partial X_j} + \frac{\partial U_j}{\partial X_i}\right), \qquad (4.4.7)$$

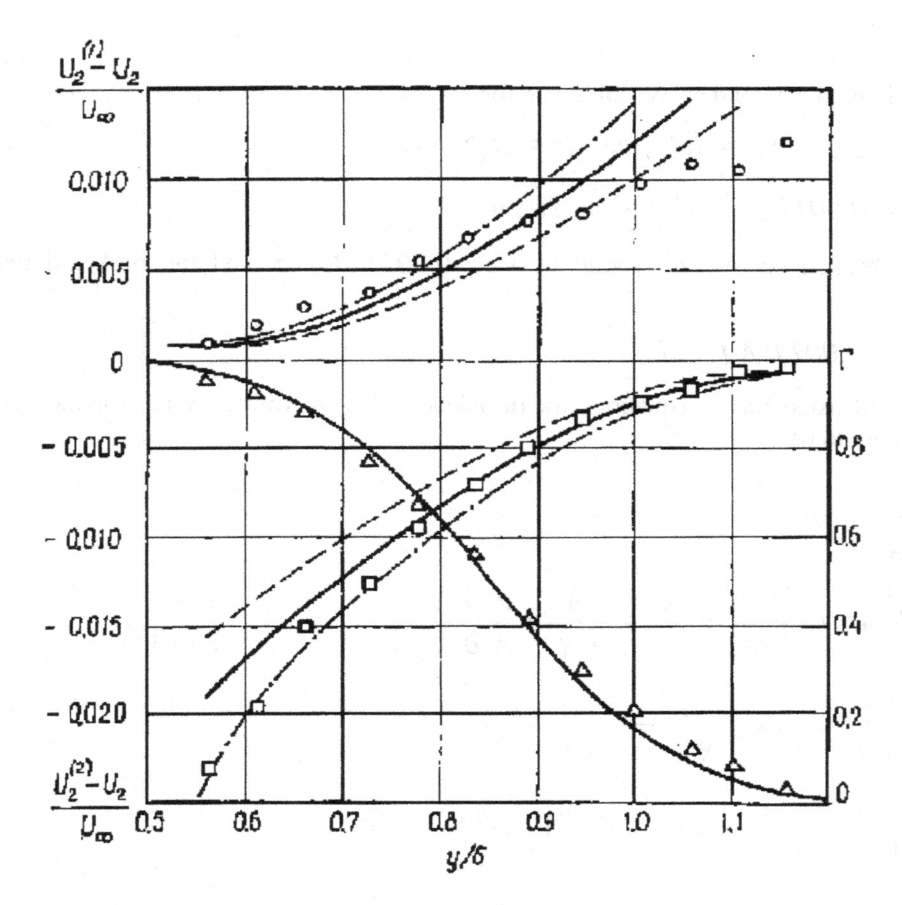

Figure 4.4.1. Comparison of formula (4.4.4) with experimental data [145] for the stream out of the boundary layer at a streamlined plate

The bulk source (sink) of turbulent phase is taken as

$$\kappa = \beta \Gamma \rho. \tag{4.4.8}$$

Then evolution equation (4.4.6) for the coefficient of intermittency has a form analogous to the transfer equations for "density of turbulence"(see [256]).

In the local-stationary state $\beta = 0$ and the coefficient of intermittency is equal to its equilibrium value $\Gamma = \Gamma_0 = const$. In general case we represent the β quantity as:

$$\beta = \beta_0 \frac{\nu}{\lambda^2}(\Gamma_0 - \Gamma), \tag{4.4.9}$$

Here λ is on the scale of a turbulent spot. In general, the Γ_0 level, determining the turbulence level, depends on the gradient of mean velocity so long as the turbulence is supported by an energy supply from a mean flow with a gradient of mean velocity unequal to zero. The resulting equation for the intermittency coefficient has the form

$$\frac{\partial \Gamma}{\partial t} + U_j \frac{\partial \Gamma}{\partial X_j} = \frac{\beta_0 \nu}{\rho l^2}(\Gamma_0 - \Gamma)\Gamma + \frac{\partial}{\partial X_j}\left(\frac{\nu}{A}\frac{\partial \Gamma}{\partial X_j}\right). \tag{4.4.10}$$

We shall write equations (4.4.6) in approximation of the boundary layer theory using (4.4.7) and (4.4.10):

$$\frac{\partial U_1}{\partial X_1} + \frac{\partial U_2}{\partial X_2} = 0,$$

$$\frac{\partial \Gamma}{\partial t} + U_1 \frac{\partial \Gamma}{\partial X_1} + U_2 \frac{\partial \Gamma}{\partial X_2} = \frac{\beta_0 \nu}{\rho \lambda^2}\Gamma(\Gamma_0 - \Gamma) + \frac{\partial}{\partial X_2}\left(\frac{\nu}{A}\frac{\partial \Gamma}{\partial X_2}\right), \tag{4.4.11}$$

$$\frac{\partial U_1}{\partial t} + U_1 \frac{\partial U_1}{\partial X_1} + U_2 \frac{\partial U_1}{\partial X_2} = \frac{\partial}{\partial X_2}\left\{[\nu\Gamma + \nu_0(1 - \Gamma)]\frac{\partial U_1}{\partial X_2}\right\}.$$

The experimental investigation [32] of intermittency inside the boundary layer of a streamlined plate showed that a tangential stress is mainly determined by the term $\Gamma\nu(\partial U_i/\partial X_j + \partial U_j/\partial X_i)$, i.e. the following inequality is valid:

$$\nu\Gamma >> \nu_0(1 - \Gamma). \tag{4.4.12}$$

If we assume that the turbulent viscosity is constant for the considered flows [110, 144, 281], then equations (4.4.11) in the Oseen approximation for the present wake are simpler:

$$U_\infty \frac{\partial \Gamma}{\partial x} = \frac{\beta_0 \nu}{\rho l^2} \Gamma (\Gamma_0 - \Gamma) + \frac{\nu}{A} \frac{1}{y^n} \frac{\partial}{\partial y} (y^n \frac{\partial \Gamma}{\partial y}) \qquad (4.4.13)$$

,

$$U_\infty \frac{\partial U}{\partial x} = \nu \frac{1}{y^n} \frac{\partial}{\partial y} (y^n \Gamma \frac{\partial U}{\partial y}).$$

Here $U = U_1 - U_\infty$ is the velocity defect, U_∞ is a velocity of external flow, x is a distance from the body downwards the flow, y is a cross coordinate. The case $n = 0$ conforms to a plane wake, but $n = 1$ conforms to a wake of a body of revolution.

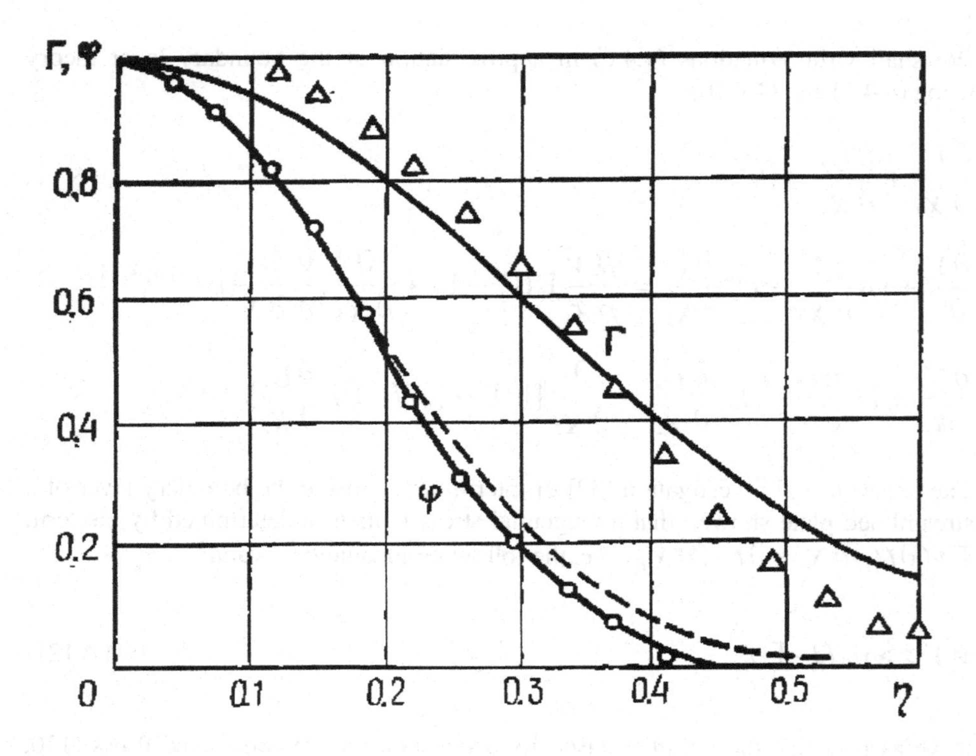

Figure 4.4.2. Plane wake behind a cylinder: lateral velocity profile and intermittency

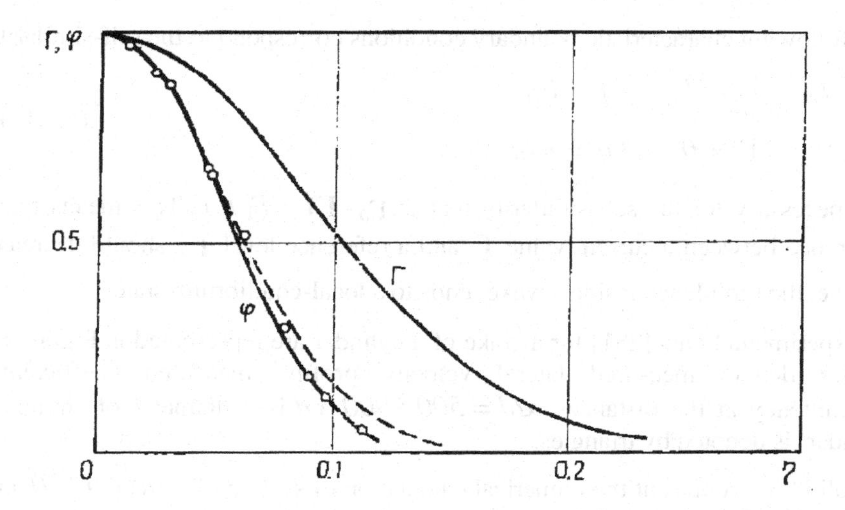

Figure 4.4.3. Wake behind a body of revolution: lateral profile of velocity and intermittency

The experiment [110, 281] shows that the axial velocity deviation from the average field is decreasing as $x^{-1/2}$ in the plane case and as x^{-1} in the axisymmetric case. The similarity conditions for the lateral distribution of axial velocity

$$\varphi = \frac{U}{U_{\max}}$$

as well as of intermittency coefficient are also satisfied, and this indicates the self - similarity of the whole problem. Introduction of a corresponding variable

$$\eta = \frac{y}{\sqrt{x\, d}}\,,$$

where d is a characteristic length, transforms equations (4.4.13) into the form

$$\begin{aligned} -\eta \frac{d\Gamma}{d\eta} &= \frac{\beta_0 \nu}{\rho \lambda^2} \Gamma (\Gamma_0 - \Gamma) x + \frac{2\nu}{A U_\infty d} \frac{1}{\eta^n} \frac{d}{d\eta} (\eta^n \frac{d\Gamma}{d\eta}), \\ -\eta \frac{d\varphi}{d\eta} &= \varphi + \frac{2\nu}{U_\infty d} \frac{1}{\eta^n} \frac{d}{d\eta} (\eta^n \Gamma \frac{d\varphi}{d\eta}), \end{aligned} \qquad (4.4.14)$$

The following characteristic boundary conditions correspond to the self-similarity:

$$\varphi = 1, \quad \Gamma = 1 \quad (\eta = 0),$$
$$\varphi = 0, \quad \Gamma = 0 \quad (\eta = \infty). \tag{4.4.15}$$

It is necessary for the self-similarity that $x(\Gamma_0 - \Gamma) = F(\eta)$. This means that the difference between a current value Γ and a reference level Γ_0 should decrease as x^{-1}, i.e. the turbulence inside a wake tends to a local-equilibrium state.

Experimental data [281] for a wake of a cylinder are represented in Figure 4.4.2. Circles denote measured lateral velocity profile; measured coefficient of intermittency at the distances $x/d = 500 \div 900$ (d is a diameter of streamlined cylinder) is denoted by triangles.

Solid lines represent the numerical calculation of system (4.4.14) if $n = 0$ under the Cauchy boundary conditions

$$\varphi = 1, \quad \Gamma = 1, \quad \frac{d\varphi}{d\eta} = 0, \quad \frac{d\Gamma}{d\eta} = 0 \quad (\eta = 0) \tag{4.4.16}$$

and the following parameters values are selected

$$\frac{\nu}{U_\infty d} = 0.016, \qquad A = 0.36, \qquad \frac{\beta_0 \nu}{\rho \lambda^2}(\Gamma_0 - \Gamma)x = 1.$$

The calculation leads to satisfaction of condition (4.4.15) if $\eta = \infty$. The broken line s represents the distribution of velocity without accounting for intermittency.

Analogous curves for a wake behind a body of revolution are represented in Figure 4.4.3.

As it was done in the plane case, it is assumed that

$$A = 0.36, \qquad \beta_0 x = 1, \ \nu = 0.00125\, U_\infty d$$

where d is a diameter of middle cross - section.

The experimental points correspond to distances from the body equal to $x/d = 12$ and to $x/d = 18$.

CHAPTER 5

GEOPHYSICAL TURBULENCE

5.1. THE TORNADO AS A LOCALIZED PHENOMENON

The problem of hurricanes and tornadoes is one of the most intriguing geophysical phenomena and its consideration is based on a nonlinear variant of the theory under consideration. Possible dependency of turbulent viscosities on spin eddy rotation rate leads to nonlinear diffusion equations and this means the localization of turbulent object by a front, slowly spreading in the atmosphere. Two variants are considered - with constant [119] and changing eddy moment of inertia [120]. The latter gives the solution with internal core ("eye") rotating in a solid manner. The coincidence with hurricane David is shown.

Angular momentum always exerts a stabilizing effect on rotating objects and it cannot be omitted when the tornado theory is being considered [9]. Kinematical independent spin velocity of an eddy (in addition to mean rotation rate of a flow) permits us to account for the rotation inertia connected with the eddy size. Of course, the inertia eddy moment changes due to the turbulent flow evolution. The tornado belongs to this type of phenomena where acting stresses possess asymmetric properties.

We accept the given above thermodynamic analysis of a turbulent object as an "open" system (Chapter 3) and accordingly turbulent viscosity is a function of the energy flux, this time expressed through the eddy spin velocity. The latter is invariant to the rigid translation and rotation of the coordinate system and can play the role of thermodynamic parameter.

Some well-known studies of tornadoes [147, 273, 298, 299] were based on the Navier-Stokes equations that permits only smooth solutions, from which one can suggest fitting of piece-wise smooth solutions.

However, turbulent viscosity, depending on spin velocity, may be used to avoid such situations by adding the effects of the angular momentum. In the zones of high gradients the front can appear and an initial discontinuity does not disappear instantly due to the effects of essentially nonlinear diffusion. In other words, we get the opportunity to model localized turbulent rotating objects.

Let us consider the corresponding equations of the turbulence with eddy angular momentum. The first is the conventional mass balance

$$\nabla_i U_i = 0 \tag{5.1.1}$$

where U_i is the flow velocity. The impulse balance is the second one:

$$\frac{\partial U_i}{\partial t} + \left(U_j \nabla_j\right) U_i = -\frac{1}{\rho} \nabla_i p + (\nabla_j \nu \nabla_j) U_i - 2\nabla_j (\gamma \varepsilon_{ijk} \omega_k) \tag{5.1.2}$$

with ρ - density, p - pressure, ω_i - eddy spin, $\nabla_i \equiv \partial / \partial x_i$- the Hamilton operator, ν, η, γ- kinematics' turbulent viscosity.

The angular momentum balance has to be added to the previous one and has the following form

$$\begin{gathered} \frac{\partial}{\partial t}(\Omega_i + \omega_i) + (U_j \nabla_j)(\Omega_i + \omega_i) = \\ (\nabla_j 2\eta \nabla_j)(\Omega_i + \omega_i) - \frac{4\gamma}{J}\omega_i \end{gathered} \tag{5.1.3}$$

Here Ω_i is average vorticity; J is the scaled eddy inertia moment (here constant); the simplest estimation follows from the assumption that a small turbulent eddy rotates with spin velocity ω_i as a solid core in a vortex line - due to high viscous stresses, γ is the turbulent rotation viscosity.

Turbulent viscosities (shear and couple) are assumed as follows:

$$\nu = \nu_* \omega^n, \quad \eta = \eta_* \omega^n \tag{5.1.4}$$

where ν_*, η_* are constant, $n \geq 0$.

Due to the axial symmetry, the radial velocity is zero: $U_r = 0$. Then only three unknown variables are independent:

$$\omega \equiv \omega_z(t,r), \qquad U \equiv U_\theta(t,r), \qquad p = p(t,r)$$

Then the impulse balances along radial and hoop coordinate lines give two equations *(n=0)*:

$$\frac{\partial U}{\partial t} = \frac{1}{r}\frac{\partial}{\partial r}\left(\nu r \frac{\partial U}{\partial r}\right) + 2\gamma \frac{\partial \omega}{\partial r} - \frac{U}{r^2}\frac{\partial}{\partial r}(\nu r) \tag{5.1.5}$$

$$U^2 = \frac{r}{\rho}\frac{\partial p}{\partial r} \tag{5.1.6}$$

Angular momentum balance (5.1.3) has the form:

$$\frac{\partial \omega}{\partial t} = \frac{1}{r}\frac{\partial}{\partial r}[(2\eta-\gamma) r \frac{\partial \omega}{\partial r}] - \frac{4\gamma}{J}\omega + \tag{5.1.7}$$

$$+ \frac{1}{2r}\frac{\partial}{\partial r}[(2\eta-\nu) \; r \; \frac{\partial}{\partial r}(\frac{1}{r}\frac{\partial rU}{\partial r}) - \frac{\partial \nu}{\partial r} \; r \; (\frac{\partial U}{\partial r} - \frac{U}{r})]$$

$$\Omega = \frac{1}{r}\left(\frac{\partial rU}{\partial r}\right)$$

where coordinate z is along the tornado axis. We have assumed that

$$2\eta \approx \nu, \qquad 2\eta >> \gamma, \qquad \gamma = const$$

The last term corresponds to turbulence generation and is equal to zero in the case of solid body rotation, that is, when $U \sim r$. If this term is negligible, angular momentum balance takes a form of *nonlinear spin diffusion* with a sink:

$$\frac{\partial \omega}{\partial t} = \frac{1}{r}\frac{\partial}{\partial r}[\, 2\,\eta_* \,\omega^n \,(r \frac{\partial \omega}{\partial r})\,] - \frac{4\gamma}{J}\omega \tag{5.1.8}$$

Time scale change:

$$t' = \frac{2\eta_* t}{(n+1)}$$

transforms the latter equation:

$$\frac{\partial \omega}{\partial t} = \frac{1}{r}\frac{\partial}{\partial r}(r \frac{\partial \omega^{n+1}}{\partial r}) - \frac{4\gamma (n+1)}{2\eta_* J}\omega \tag{5.1.9}$$

The substitution

$$\omega(t^{'}, r) = \Phi(t^{'}, r)\exp[-\frac{2\gamma(n+1)t^{'}}{(\eta_* J)}] \tag{5.1.10}$$

$$\tau = \frac{\eta_* J}{2\gamma n(n+1)}[1 - \exp(-\frac{4\gamma n t}{J})] .$$

yields the well-known form for the spin diffusion equation

$$\frac{\partial \Phi}{\partial \tau} = \frac{1}{r}\frac{\partial}{\partial r}(r\frac{\partial \Phi^{n+1}}{\partial r}) \tag{5.1.11}$$

with the following self-similar solution:

$$\Phi(\xi) = [\frac{Q}{4\tau}]^{\frac{1}{n+1}} [\frac{n}{4(n+1)^2}(\xi_0^2 - \xi^2)]^{\frac{1}{n}}, \quad \xi < \xi_0 \tag{5.1.12}$$

$$\Phi(\xi) = 0, \qquad \xi \geq \xi_0$$

This solution corresponds to the instant singular source of intensity Q in a resting atmosphere:

$$\omega = 0, \ r \Rightarrow \infty; \omega = Q_\omega \delta(r^2), \ t = 0 \tag{5.1.13}$$

where $\delta(r)$ – the Dirac delta function.

The nonlinear diffusion corresponds to *the front of a large turbulent vortex, widening with finite velocity in the radial direction.* Outside there is no turbulence at all.

$$\xi = r[\tau(\frac{Q}{4})^n]^{\frac{1}{2(n+1)}},$$

$$\xi_0 = [\frac{4(n+1)^2}{n} 2^n \left(\frac{n+1}{n}\right)^n]^{\frac{1}{2(n+1)}} . \tag{5.1.14}$$

Therefore, the *front radius* is determined by equality $\xi = \xi_0$, that is,

$$R = \left[\tau \frac{4\,(n+1)^2}{n} \frac{Q^n\,(n+1)^n}{2^n\,n^n} \right]^{\frac{1}{2(n+1)}} \tag{5.1.15}$$

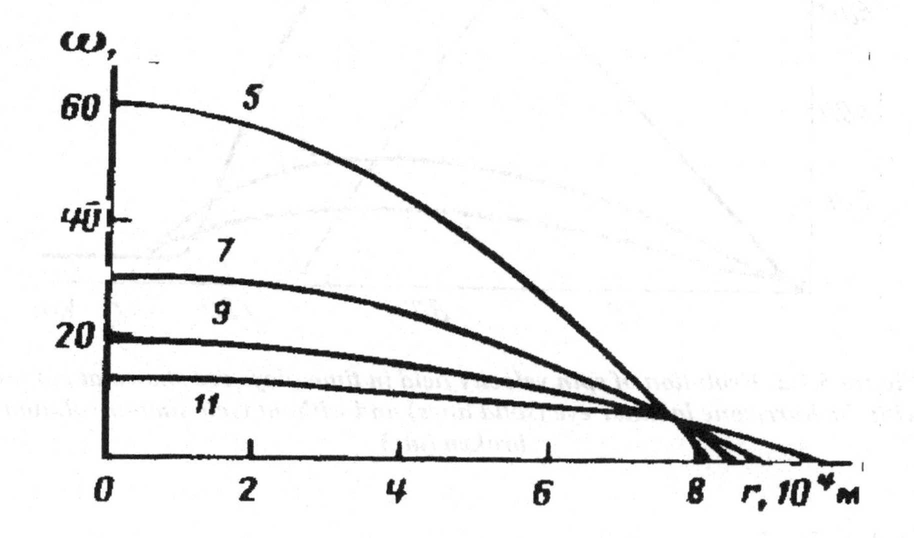

Figure 5.1.1. Growth of hurricane outer boundary (ω in 1/s)

The considered turbulent object is always localized because:

$$\tau \to \tau_{\infty} = \frac{\eta_* J}{2\,\gamma\,n\,(n+1)} < \infty \qquad (t \to \infty)$$

The corresponding curves are given in Figure 5.1.1 for days of a hurricane's life.

This effect is quite similar mathematically to the "fireball" in the atmosphere: see [17, 304, 305]. The first who mentioned this property of the nonlinear diffusion was Mrs. P. Y. Kochina [238].

Now we consider *the case of the inertia moment evolution*. The corresponding equation for turbulent eddy inertia moment has the form:

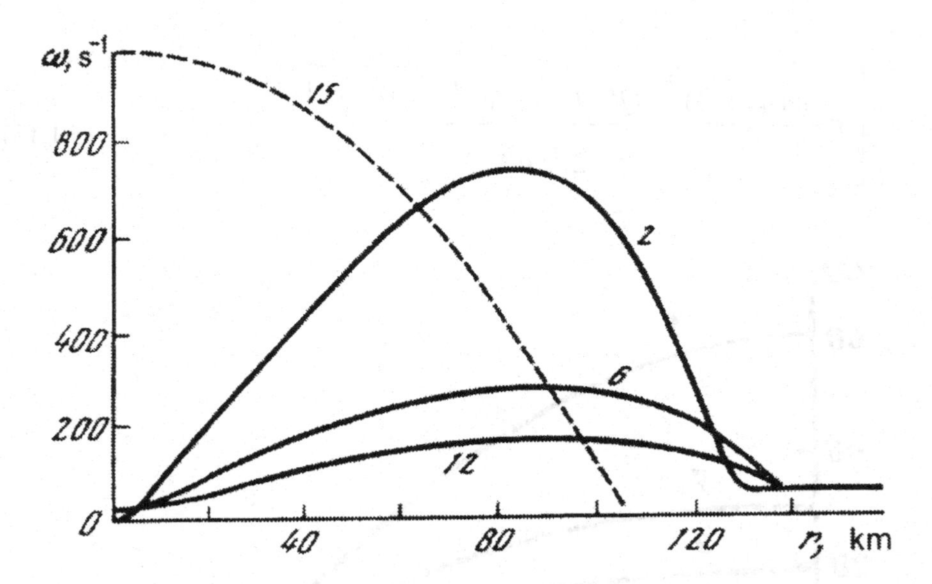

Figure 5.1.2. Evolution of spin velocity field in time (days are shown at curves) with the hurricane laminar eye (solid lines) and without (self-similar solution - broken line)

$$\frac{\partial J}{\partial t} = \frac{1}{r}\frac{\partial}{\partial r}\left(\varsigma \; r \frac{\partial J}{\partial r}\right) \tag{5.1.16}$$

We assume that $\varsigma = 2\eta$. Accounting for this effect needs in a more complicated angular momentum balance:

$$\frac{\partial \omega}{\partial t} = \frac{1}{r}\frac{\partial}{\partial r}\left[(2\eta - \gamma)\, r \frac{\partial \omega}{\partial r}\right] - \frac{4\gamma}{J}\omega$$

$$+ \frac{1}{r}\frac{\partial}{\partial r}\left[(2\eta - \nu)\, r \frac{\partial \Omega}{\partial r} - \frac{r}{2}\frac{\partial \nu}{\partial r}\left(\frac{\partial U}{\partial r}\right.\right. \tag{5.1.17}$$

$$\left.\left. - \frac{U}{r}\right)\right] + \frac{4\eta}{J}\frac{\partial J}{\partial r}\frac{\partial}{\partial r}(\Omega + \omega)$$

The system of impulse balances and the latter two equations corresponds now to four unknown variables:

$\omega(t,r), v(t,r), J(t,r), p(t,r)$.

Because the problem became more complicated, we have to use computing methods to get the sample solution. The following initial and boundary conditions:

$$\frac{\partial \omega}{\partial r} = 0, \quad \frac{\partial U}{\partial r} = 0, \quad \frac{\partial J}{\partial r} = 0 \quad (r = 0), \tag{5.1.18}$$

$$\omega = \omega_0, \; U = 0, \; J = J_0, \quad (r \to \infty; t \geq 0) \tag{5.1.19}$$

$$\omega = 6 \cdot 10^3 \; s^{-1}, \; U = 1{,}25 \times 10^{-3} \; r,$$

$$J = 10^{-7} \; r^2, \quad (r < 10^5 \; m);$$

$$\omega = \omega_0 = 50 \; s^{-1}, \; U = 0, \tag{5.1.20}$$

$$J = J_0 = 10^3 \; m^2, \quad (r \geq 10^5 \; m, \; t = 0).$$

are assumed to be realistic. The chosen initial inertia moment corresponds to the initial eddy diameter $d = 44.7 \; m$ if

$J = \lambda^2 / 2$.

Besides we use the following values that may correspond to hurricane DAVID:

$$n = 1, \; \nu_* = 1 \; m^2, \; \eta_* = 0.5 \; m^2, \; \gamma_* = 10^{-6} \; m^2, \; \rho = 1.29 \; kg / m^3$$

The most important effect of the *J*-evolution is the hurricane's "eye" or core appearance (for some hours) as it follows from the numerical calculations.

Inside the core there is practically no turbulent transfer. Its outside boundary R_0 is determined by inequality $\omega_0 \leq \omega$. The "eye" grows with a speed of 1.6 km/day and its radius $R_0 = 20$ km after 20 days. Inside, the "solid body rotation" means $U \sim r$.

The external radius R of a turbulent hurricane is stable and equal to *140 km* after *4 days*. At R_m~*60 km* the wind (tangential) velocity has maximum equal to 75 m/s. The wind decreases as $r^{-\chi}$ at $r > R_m$, where $\chi = 0,5 - 1$. According to J – data, the individual turbulent eddy diameter changes from a few meters to tens of meters along the radius of the hurricane.

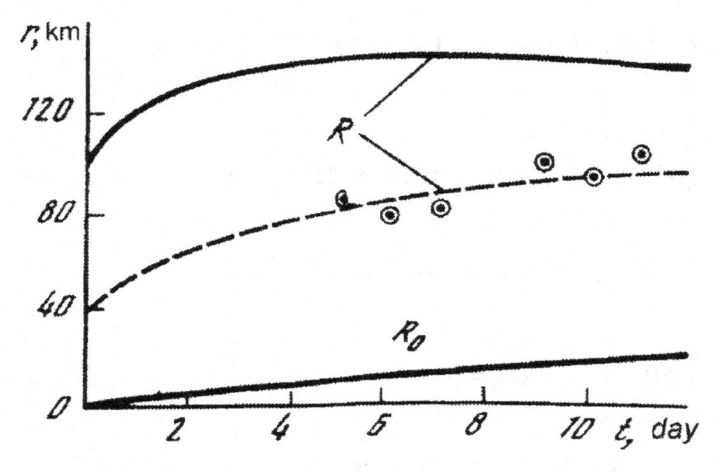

Figure 5.1.3. External R and internal R_0 boundaries of the hurricane turbulent zone (the broken line corresponds to the self-similar solution without "eye")

The coincidence with DAVID's data[19] shows that this hurricane had no "eye". The scale of our solution is in accordance with hurricane data.

A better coincidence with a *tornado*, that is much smaller, will be reached for bigger density ρ and inertia ρJ to include the suspended water drops or dust particles.

Correspondingly, the effect of tornado destabilization will be reached if these suspended particles can be deleted. Moreover, it is quite possible to think that real *tornadoes are stable only because of suspended heavy additives.*

---

[19] Bencloski T.W. J. Geogr. # 9, 204 - 219 (1981); Willoughby H.E., Clos T.A., Shoreibak M.G., J. Atmos. Sci., v. 29, #3, 395-411 (1982).

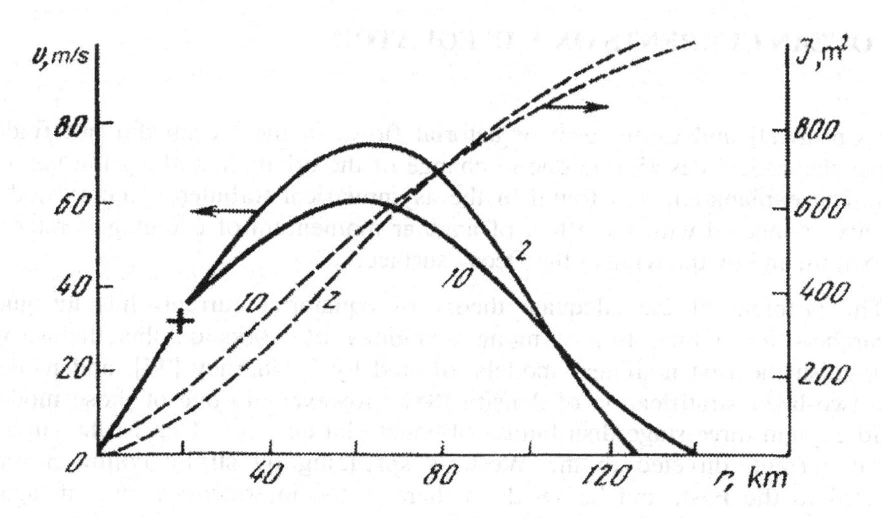

Figure 5.1.4. Velocity U and inertia moment J inside the hurricane with "eye": the eye radius R_0 is shown by cross-symbol.

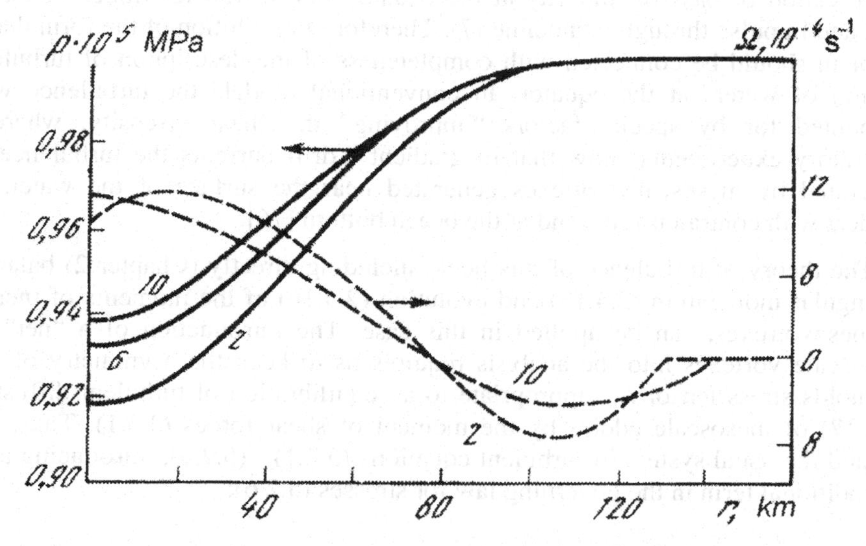

Figure 5.1.5. The hurricane with "eye" is shown. Pressure is decreasing dangerously to the centre. Mean vortex Ω changes its sign as it was seen in laboratory experiments [35]

5.2. OCEAN CURRENTS ON THE EQUATOR

The Cromwell and Lomonosov equatorial flows in the Ocean did not find a proper theoretical description due to change of the orientation along the vertical axis. The explanation was found in the asymmetrical turbulence theory and is directly connected with the effect of angular momentum of eddies generated at the bottom and by the wind at the Ocean surface.

The problem of the adequate theory of equatorial currents has intrigued researchers for a long time. Among a number of works on this theme, we distinguish the first nonlinear models, offered by J. Charney [44], and models with two-layer stratification of density [85]. However, no one of those models could explain three-stage distribution of equatorial currents. That is, the surface passat current, directed to the West, is spreading by strong contra-current, directed to the East, and below them there is the intermediate current, again directed to the West and settled down in the main *pycnocline*.

First of all, observations have shown, that currents at the equator are essentially turbulent [7, 84, 85]. It can be seen, in particular, in the scatter of the stratification of passive impurity at the equator and in transfer fluxes of mass, heat and impulse through picnocline [7]. Therefore the solution of the formulated problem should be connected with completeness of the description of turbulent motion of waters at the equator. In conventional models the turbulence was accounted for by special factors "improving" the shear viscosity, whereas laboratory experiments show that in gradient - drift currents the turbulence is generated by mesoscale vortexes generated near the surface of the water, at borders with contra-currents and at the ocean bottom [76].

The theory of turbulence of this book, including directly (Chapter 2) balance of angular momentum (2.4.13) and evolution (2.4.16) of the moments of inertia of mesovotrexes, can be applied in this case. The introduction of a "net" of mesoscale vortexes into the analysis requires us to keep the asymmetry of the Reynolds stress tensor R_{ij}, appropriate to an equilibration of turbulent diffusion (2.3.17) of mesoscale eddies by the moment of shear forces (3.3.1). Thus, we expand the usual system of turbulent equations (5.2.1) - (5.2.5), introducing also an additional term in the governing law for stresses (5.2.6):

$$\rho \frac{\partial J \Phi_i}{\partial t} + \rho \frac{\partial}{\partial X_j} (J \Phi_i U_j) = \frac{1}{2} (\varepsilon_{ikj} R_{kj} + \frac{\partial}{\partial X_j} \mu_{ij}) \qquad (5.2.1)$$

$$\frac{\partial J}{\partial t} + \frac{\partial}{\partial X_j}(J U_j) = \frac{\partial}{\partial X_j}(\varsigma \frac{\partial J}{\partial X_j}) \tag{5.2.2}$$

$$\mu_{ij} = -4 \eta \rho J \frac{\partial \Phi_i}{\partial X_j} + \frac{\partial}{\partial X_j}(\varsigma \frac{\partial J}{\partial X_j}) \tag{5.2.3}$$

$$\rho \frac{\partial U_i}{\partial t} + \rho \frac{\partial}{\partial X_j}(U_i U_j) = -\frac{\partial p}{\partial X_j} + \frac{\partial}{\partial X_j} R_{ij} + F_i \tag{5.2.4}$$

$$\frac{\partial U_i}{\partial X_i} = 0 \tag{5.2.5}$$

$$R_{ij} = \nu_{ijkl}\, \rho (\frac{\partial U_l}{\partial X_k} + \frac{\partial U_k}{\partial X_l}) - 2 \gamma_{ijkl}\, \rho\, \varepsilon_{klm}\, \omega_m \tag{5.2.6}$$

Here $\Phi_i = -\frac{1}{2}(\Omega_i + \omega_i)$ is the total angular velocity,

$$\Omega_i = \varepsilon_{ijk} \frac{\partial U_j}{\partial X_k}$$

is vorticity of the average flow, ω_i*-spin of meso-eddies*, ε_{ijk}-Levi-Civita alternating tensor. The tensor of shear viscosity coefficients A_{ijkl} is considered to depend on the vertical direction [7]. The tensor of rotational viscosity is assumed to be isotropic:

$$\gamma_{ijkl} = \gamma\, \delta_{ik}\, \delta_{jl}$$

Let us direct the axis x to the East, the axis y to the North and the axis z downwards from the free surface of the Ocean (at $z = 0$). To study the currents in the equatorial plane (x, z), arising under action of a zonal inclination of the Ocean surface level $h(x, y)$ where g is gravity acceleration:

$$F_z = \rho g \frac{\partial h}{\partial x} = \rho g \kappa$$

we assume [7] the motion as horizontally homogeneous, that is,

$$\frac{\partial}{\partial x} = \frac{\partial}{\partial y} = 0$$

Then system (5.2.1) - (5.2.6) is reduced to the following five equations:

$$\frac{\partial}{\partial z}\left(\nu \frac{\partial U}{\partial z} + 2\gamma\omega_y\right) = g\frac{\partial \kappa}{\partial x}, \tag{5.2.7}$$

$$\frac{\partial}{\partial z}\left(2\eta J \frac{\partial \omega_x}{\partial z}\right) + \frac{\partial}{\partial z}\left(2\varsigma\ \omega_x \frac{\partial J}{\partial z}\right) = 4\gamma\omega_x, \tag{5.2.8}$$

$$\frac{\partial}{\partial z}\left[2\eta J \frac{\partial}{\partial z}(\Omega_y + \omega_y) + 2\varsigma(\Omega_y + \omega_y \frac{\partial J}{\partial z})\right] = 4\gamma\omega_y \tag{5.2.9}$$

$$\frac{\partial}{\partial z}\left(2\eta J \frac{\partial \omega_z}{\partial z}\right) + \frac{\partial}{\partial z}\left(\varsigma\omega_z \frac{\partial J}{\partial z}\right) = 4\gamma\omega_z, \tag{5.2.10}$$

$$\frac{\partial}{\partial z}\left(\varsigma \frac{\partial J}{\partial z}\right) = 0 \tag{5.2.11}$$

Here zonal velocity U, three components of the meso-eddy angular velocity and reduced inertia moment $J = \lambda^2/2$, where λ-radius of a meso-eddy, are introduced. Among three components of the vorticity Ω_i only one meridian component is not equal to zero:

$$\Omega_y = \frac{\partial U}{\partial z}$$

Equation (5.2.11) yields the first integral

$$J = J_0 + (J_H - J_0)\frac{z}{H} \tag{5.2.12}$$

where H is the depth of the Ocean layer under consideration. Solution (5.2.12) defines the character of change of the mesoeddies sizes along the vertical. However, experiments [78] show that due to spin rotation, mesoeddies are rather stable, and the sizes do not vary with depth. Hence, we can believe that

$$J_0 = J_H = J = const$$

The remaining four equations (5.2.7) - (5.2.10) have the following solution:

$$\omega_p = \omega_p^H \frac{sh(\beta z)}{sh(\beta H)} + \omega_p^0 \frac{sh[\beta(H-z)]}{sh(\beta H)} \tag{5.2.13}$$

$$\omega_y = \frac{1}{2\gamma}[\nu k \frac{U_0 sh(kz)}{ch(kH)-1} + 2\gamma\omega_y^0 ch(kz)] \tag{5.2.14}$$

$$U = \frac{g\kappa}{2\nu}(\alpha H^2 - z^2) + \frac{\tau^0}{\nu}(\alpha H - z) + \frac{2\gamma\omega_y^0}{\nu k}[\alpha\ sh(kH) - sh(kz)] + U^0 \frac{ch(kH) - ch(kz)}{ch(kH) - 1}; \tag{5.2.15}$$

where index p accepts two values (x or z) and

$$k^2 = \frac{2\nu}{\eta J[(\nu/\gamma) - 1]}, \qquad \beta^2 = \frac{2\gamma}{\eta J} \tag{5.2.16}$$

$$\alpha = \frac{ch(kz) - 1}{ch(kH) - 1}$$

$$U = U^0, \qquad -\nu\frac{\partial U}{\partial z} - 2\gamma\omega_y^0 = \tau^0,$$

$$\omega = \omega_i^0 \quad (z = 0)$$

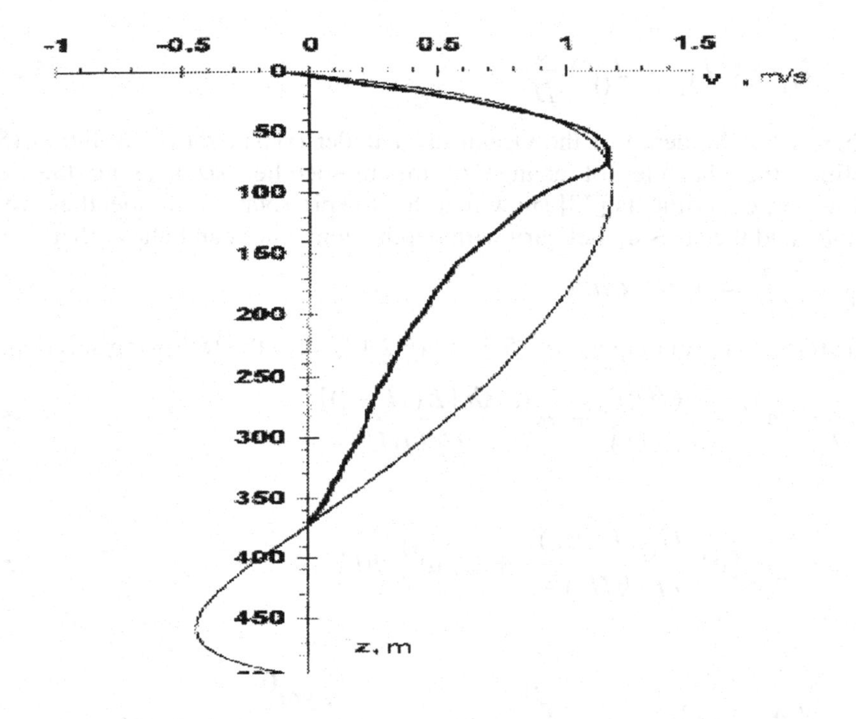

Figure 5.2.1. The Lomonosov current: Atlantic Ocean (latitude 0^0, west longitude 20-30^0). Thick line was measured. Thin line represents calculations.

$$U = 0,\ \omega = \omega_x^H, \qquad \omega = \omega_z^H \ (z = H)$$

The surface velocity U^0 and tangential wind force τ^0 are connected with zonal wind velocity *W*, see [7], by the following formulas:

$$U^0 = K\,W\ ,\ \tau^0 = c\,W^2, \qquad K = 0.02\ ,\ c = 2\times 10^{-6}$$

Detailed calculations are carried out for the Atlantic and Pacific Oceans (Figures 5.2.1, 5.2.3). The observed data of the *Lomonosov current* were used [7] as well as the results of measurements of the *Cromwell current*. They were obtained at the 17th log of "Academician Kurchatov" in the winter of 1974. As we can see, all basic features, including the maximal velocity and the currents borders and also the multistage vertical structure are really described by the theory.

The calculations were done for $\nu = 0.01\ m^2/s$, $\gamma = 0.05\ m^2/s$. Other data are given in Table 5.2.1.

Table 5.2.1.

Figure #	H m	η m^2/s	J m^2	λ m	$\gamma 10^{-8}$ m^2/s	U^0 m/s	ω_y^0 1/s	$\rho\tau^0$ Dyn/cm^2
5.2.1	500	0.011	1500	60	2	- 0.1	- 0.048	0.05
5.2.2	440	0.011	900	40	4.7	- 0.3	- 0.052	0.02
5.2.3	600	0.012	2500	70	2,7	- 0.3	- 0.044	0.16

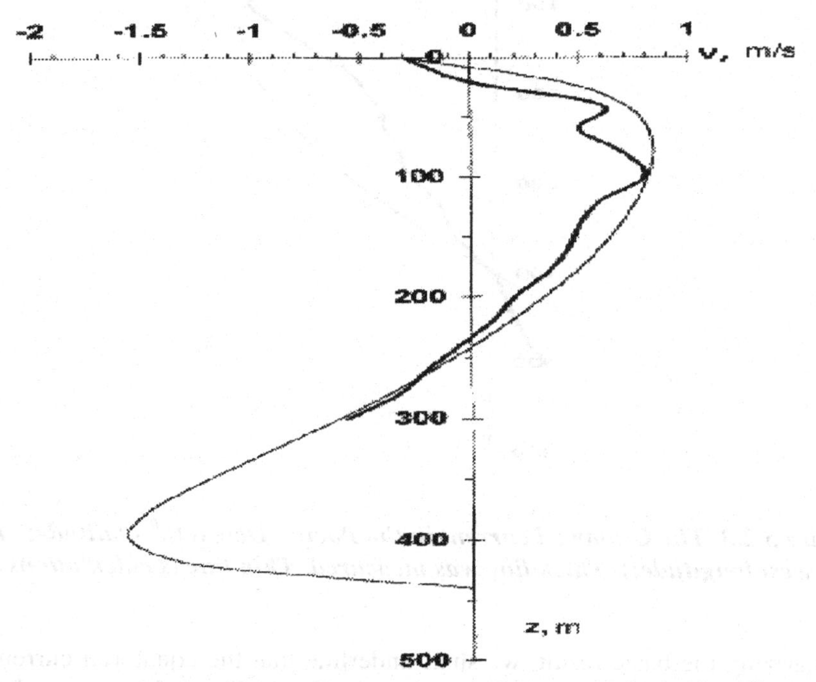

Figure 5.2.2. The Cromwell current, Pacific Ocean (latitude: 0^0; west longitude: 122^0) Thick line was measured. Thin line is calculations.

Numerical experiments show that at the passat change to the Western winds, reduction and change of signs of the Ocean surface inclination γ, change of a direction of the eddy's rotation, the flows near to a surface of the Ocean turn to the East, and the Cromwell current turns to the West, adjoining with the Western intermediate current. At the further increase of an abnormal inclination of a level γ, in the main pycnocline at the depths of 300-700 meters, the powerful intermediate current, directed to the East, is formed. The similar pattern was observed in 1982-1983 during development of the El-Nino phenomenon [84]. Thus, the intensity and directions of the currents at the equator can serve as parameters of this catastrophic phenomenon influencing the climate and circulation of waters of the Pacific Ocean and its coasts.

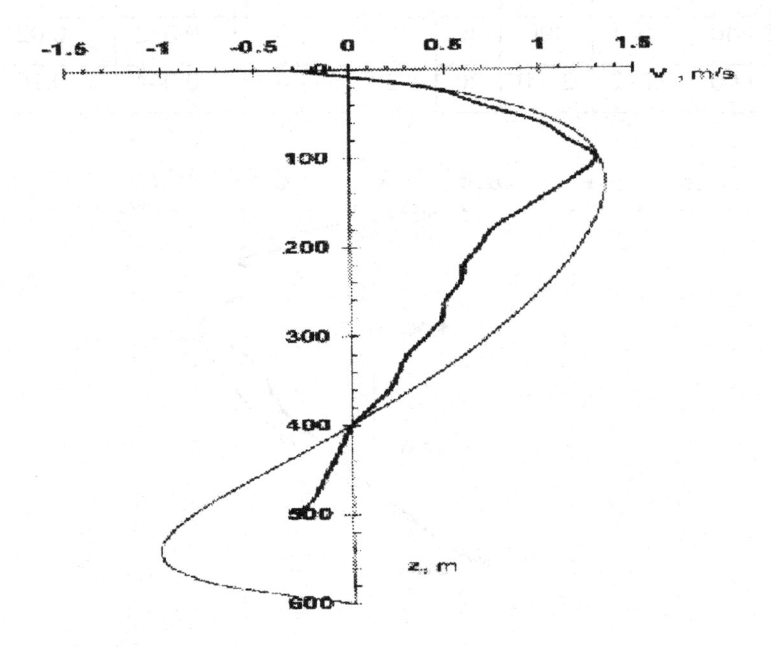

Figure 5.2.3. The Cromwell current in the Pacific Ocean (0^0 - latitude;140^0 – west longitude;). Thick line was measured. Thin line is calculations

Summarizing the basic result, we shall underline that the equatorial currents are of gradient-drift origin, containing mesoscale eddies with sizes (by our calculations) from 30 up to 50 meters. According to [229], one of the main maxims of the generalized spectra of the Ocean turbulence corresponds only to the range of scales from 7 up to 70 meters. Here there is one of three zones of power supply of the Ocean. The mesoscale eddies redistribute turbulent stresses along the vertical, forming the structure of the equatorial currents.

5.3. GLOBAL OCEAN CIRCULATION

Circulation in the world's oceans can be numerically computed but the large linear scale of the effective grid is much bigger than the *synoptic eddies* radius. This makes the situation analogous to plane turbulence but on the rotating Earth surface. The solution gives more detailed global circulation than usual ones, with rates closer to experimental data and with more distinct local circulation centers.

It is well determined now that quasi-two-dimensional synoptic eddies with a radius ~ *100 km* exist in the World Ocean [34, 193]. These eddies can essentially affect the global circulation of the ocean due to a high concentration of contained kinetic energy. Large-scale numerical calculations are necessary to reveal this effect if the spatial cell exceeds the linear scale of the synoptic eddies. Then one can use the model of turbulent fluid under consideration; the synoptic eddies playing the role of mesoscale structure having its own spin velocity ω [121, 122]. It was developed in a series of works, including ones written with participation of the present author [11, 18, 61, 62, 294].

It is known that the equations of geophysical turbulence are developed for the revolving β-plane. The momentum equation differs from [121] by the account for the Coriolis forces, that is,

$$\frac{\partial U_i}{\partial t} + \frac{\partial U_i U_j}{\partial X_j} + f\, \varepsilon_{i3k}\, U_k = -\frac{1}{\rho}\frac{\partial p}{\partial X_i} + \frac{\partial R_{ij}}{\partial X_j}, \tag{5.3.1}$$

Here $f = 2\,\Omega_3^0$ is the Coriolis parameter. Total angular momentum balance can be directly used in this case:

$$\rho J\left(\frac{\partial \omega_3}{\partial t} + \frac{\partial \omega_3 U_j}{\partial X_j}\right) + \rho\,(2\kappa^2 + J)\left(\frac{\partial \Omega_3}{\partial t} + \frac{\partial \Omega_3 U_j}{\partial X_j} + \right.$$
$$\left. + \Omega_3^0 \frac{\partial U_3}{\partial X_3} + \frac{\beta}{2} U_2\right) = \frac{\rho J}{2}\,\beta\, U_2 + \frac{\partial \mu_{3j}}{\partial X_j} - \varepsilon_{3lk} R_{lk} \tag{5.3.2}$$

$$\beta = 2\frac{\partial \Omega_3^0}{\partial X_2}$$

Here β is the alternation of the Carioles parameter with latitude (this is the so-called β-effect), Ω_3^0 is the angular velocity of the Earth's rotation ($= 7{,}29 \times 10^{-5}\ s^{-1}$) and κ is the average scale of the spatial averaging:

$$<\rho> J_{ml} =<\rho \xi_m \xi_l>=<\rho> \lambda_L^2 \delta ml \qquad (m,l = 1,2)$$

$$\kappa_L^2 = L^2/12 \qquad \kappa_H^2 = H^2/12 = \kappa^2; \tag{5.3.3}$$

$$\frac{\partial \omega_3}{\partial t} + \frac{\partial \omega_3 U_j}{\partial X_j} = -\frac{\phi}{2}\frac{\partial}{\partial X_j}\left(\frac{\partial R_{2j}}{\partial X_1} - \frac{\partial R_{1J}}{\partial X_2}\right) +$$

$$\frac{\beta}{2} U_2 + \frac{\eta}{J}\Delta^2(\Omega_3 + \omega_3) - \frac{1}{J}\varepsilon_{3lk} R_{lk}, \tag{5.3.4}$$

$$\phi = 2\frac{\kappa^2}{J} + 1.$$

The spherical coordinate system will be used below instead of the Cartesian one, the projections of the velocity vector being correspondingly designated as l, θ, z. We average equation (5.3.1) and (5.3.2) over the vertical coordinate X_3. It leads to the inclusion of components of turbulent friction spatially averaged forces τ_i^H, τ_i^0, at the surface and bottom of the ocean. Besides, the stream function ψ is introduced (now for mean velocities U_θ, U_l,) by the rule

$$U_l = \frac{1}{H}\frac{\partial \psi}{\partial \theta}, \qquad U_\theta = -\frac{1}{H \sin\theta}\frac{\partial \psi}{\partial l} \tag{5.3.5}$$

Here H is the Ocean depth. Now impulse balance (2.2.14) will have the form:

$$\frac{\partial \Delta_H \psi}{\partial t} - \frac{1}{R}\left[\frac{\partial}{\partial \theta}\left(\frac{\Delta_H \psi}{H \sin\theta}\frac{\partial \psi}{\partial l}\right) - \frac{\partial}{\partial l}\left(\frac{\Delta_H \psi}{H \sin\theta}\frac{\partial \psi}{\partial \theta}\right)\right] -$$

$$-\left[\frac{\partial}{\partial \theta}\left(\frac{f}{H}\frac{\partial \psi}{\partial l}\right) - \frac{\partial}{\partial l}\left(\frac{f}{H}\frac{\partial \psi}{\partial \theta}\right)\right] \tag{5.3.6}$$

$$= \frac{\partial}{\partial\theta}\left(\frac{\tau_l^H - \tau_l^0}{\rho H}\right) - \frac{\partial}{\partial\lambda}\left(\frac{\tau_\theta^H - \tau_\theta^0}{\rho H}\right) - \frac{2\gamma}{J}\Delta_H\,\omega,$$

where R is the Earth's radius, θ is the latitude addition to 90° , l is the longitude and the z axis is directed downwards and:

$$\omega = -\int_0^H \omega_3\, dz, \tag{5.3.7}$$

$$\Delta_H\,\psi = \frac{\partial}{\partial\theta}\frac{\sin\theta}{H}\frac{\partial\,\psi}{\partial\,\theta} + \frac{1}{\sin\theta}\frac{\partial}{\partial l}\frac{1}{H}\frac{\partial\,\psi}{\partial\,l}$$

The averaging of equation (5.3.4) over a vertical axis leads to the following equation:

$$\frac{\partial\,\omega}{\partial\,t} + \frac{1}{R\sin\theta}\left[\frac{\partial\,U_\theta\,\omega\sin\theta}{\partial\,\theta} + \frac{\partial\,U_l\,\omega}{\partial\,l}\right] =$$

$$\frac{\eta}{J\,R^2\sin\theta}\Delta^2\left(\frac{\partial\,U_l\sin\theta}{\partial\,\theta} - \frac{\partial\,U_\theta}{\partial\,l}\right)$$

$$+\frac{1}{R^2}\left(\frac{\eta\,\rho}{J} + \gamma\right)\Delta^2\,\omega + \frac{4\gamma}{J}\,\omega + \tag{5.3.8}$$

$$\frac{\beta}{2}U_\theta\,H + \frac{\phi}{2R\sin\theta}\left[\frac{\partial\,(\tau_l^H - \tau_l^0)}{\partial\,\theta} - \frac{\partial(\tau_\theta^H - \tau_\theta^0)}{\partial\,l}\right].$$

if there is no flux of ω_3 through the free surface and $\omega_3 = 0$ at the Ocean bottom. Here $r = R$, θ and l are spherical coordinates.

We approximate the World Ocean by the double - connected area (Figure 5.3.1). Its North boundary G_1 goes along the seacoast of Australia, Eurasia (as the a whole), Africa and America. Its South boundary G_2 goes along the seacoast of the Antarctic

Continent (~ 70° of South latitude). The boundary conditions are given for stream-function ψ and spin velocity ω

$$\psi = 0, \quad \omega = 0 \quad (l, \theta \in G_1), \tag{5.3.9}$$

$$\psi = C, \quad \partial\omega/\partial n = 0 \quad (l, \theta \in G_2), \tag{5.3.10}$$

Constant C is determined by the condition of periodicity along the contour

$$\oint [\frac{\partial}{\partial\theta}(\frac{1}{H}\frac{\partial\psi}{\partial\theta}) + \frac{\tau_l^H - \tau_l^0}{\rho H} + \frac{2\gamma}{H}\frac{\partial\omega}{\partial\theta}]\ dl = 0. \tag{5.3.11}$$

Condition (5.3.9) means absence of synoptic eddy generation at the contour G_1. Condition (5.3.10) means the absence of eddy flux along the normal n to the contour G_2.

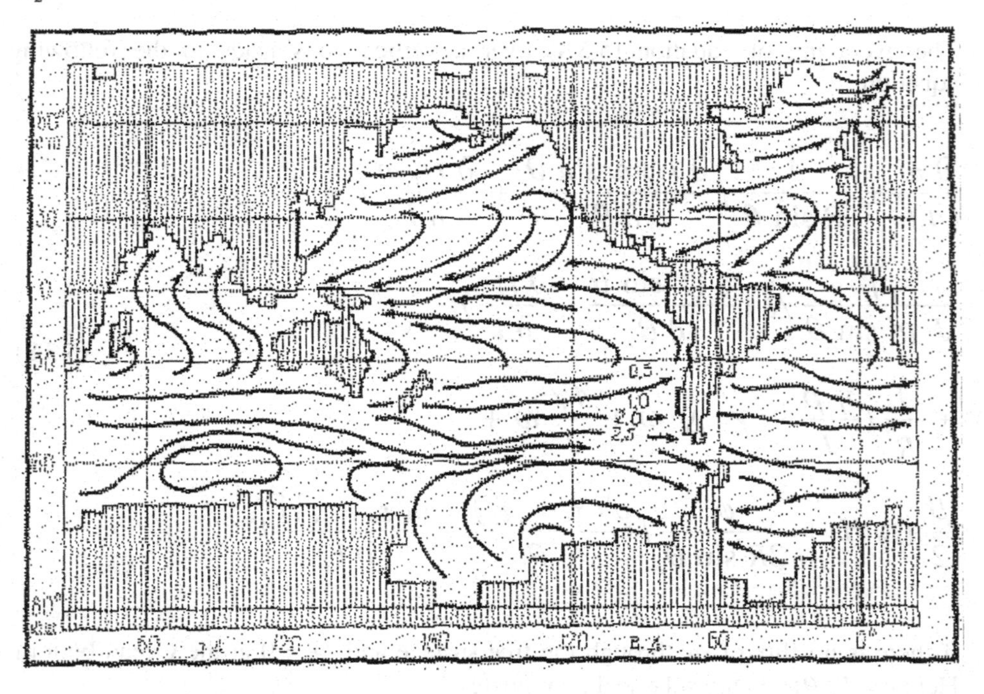

Figure 5.3.1. The contour map and winds according to the Hellerman data

The tangential forces, created by the wind τ_l^0, τ_θ^0, were given in accordance with the Hellerman wind data [108] for a typical winter season. Depths field $H(\lambda,\theta)$ was given according to the data in [232]. As to the calculation of resistance at the Ocean bottom τ_l^H, τ_θ^H, these forces were determined by the methods in [127, 151] as functions of the Rossby number

$$Ro = \frac{W}{f z_0}.$$

Here W is the velocity of flow at the upper boundary of the near-bottom boundary layer and z_0 is the bottom roughness ($\sim 0{,}01\ m$).

The numerical integration of equation (5.3.6), (5.3.7) was carried out mainly by A.I. Danilov [61, 62]. He used implicit schemes of the first order of accuracy over the time and over the horizon coordinates. The second and the third terms in the left-hand side of the equation for ψ were approximated at the upper $(k+1)$-th time layer in the same way. The last term in the right-hand side of (5.3.7) is taken at the upper time layer. The following divergent form approximates the convection of ω.

$$\frac{1}{R\sin\theta}[(Y^+(U_l)\omega)_{-l} + (Y^-(U_\lambda)\omega)_{+l}$$

$$+ (Y^+(U_\theta\sin\theta)\omega)_{-\theta} + (Y^-(U_\theta\sin\theta)\omega)_{+\theta}]. \tag{5.3.12}$$

Here the lower indexes: $+l$, $-l$, $+\theta$, $-\theta$, denote the direction of the difference: $+l$ is forward, $-l$ is backwards,

$$Y^+(a) = \begin{cases} a, & a > 0 \\ 0, & a \le 0, \end{cases}$$

$$Y^-(a) = \begin{cases} 0, & a > 0, \\ a, & a \le 0. \end{cases} \tag{5.3.13}$$

The terms in (5.3.7), containing ψ, were taken at the time level $k+1$. At the initial time, the ocean was considered to be at rest. The adjustment of ψ and ω fields was realized at each time step, that is, ω_{p+1}^{k+1} was determined by the $(p+1)$-th approximation of the ψ^{k+1}, and the next approximation for ψ^{k+1} was calculated by the ω_{p+1}^{k+1}. This process was interrupted when the maximum of relative difference of the fields ψ_p^{k+1} and ψ_{p+1}^{k+1} did not exceed 1%. The chosen time step is twenty-four hours; the chosen space step is 5^0. The circulation was considered steady when the maximum of relative difference of ψ^{k+1} and ψ^k was less than 1%.

The distribution of the function of current (m^2 / s) in the World Ocean calculated at $\omega = 0$ is shown in Figure 5.3.2. It coincides well enough with the results in [127, 151] where a similar problem was considered. Some deviations are caused by the differences in the given fields τ_l^{0l}, τ_θ^0.

The steady average circulation of the ocean (under assumption that $\rho = const$) taken into account the synoptic eddies ($\omega \neq 0$) is represented in Figure 5.3.3. In this case, the following values are used:

$$\kappa = 10^6, \; J = 4 \times 10^{10} \; m^2, \; \eta = 10^{18} \; m^4 /s, \; \gamma = 5 \times 10^3 \; m^2 /s$$

Comparison of Figures 5.3.2 and 5.3.3 allows us to see the changes, produced by the synoptic eddies. It is curious that changes of the global circulation do not take place. The qualitative differences reduce to resolution of the epicenters of some rotation (marked with arrows and coincided partly with the Sanderson map), including the center, which can be referred to as to Bermuda triangle. It is known that the reduction of the ocean level exactly in this region, registered by observations from space, could be interpreted as a result of large vortex existence.

The tendency to flow rate increase in anticyclone rotations of the North Atlantic and in the Pacific turns one's attention to the quantitative changes. Thus, the Gulf Stream rate has increased from $18{,}5 \times 10^6 \; m^3 / s$ up to $22{,}1 \times 10^6 \; m^3 / s$ (the real one is $80 \cdot 10^6 \; m^3 / s$). The Kurosio has increased from 35 up to 40,1 million m^3 / s (the real one is ~ 50), whereas the flow rate of the Antarctic Circumpolar Current (ACC) - $260 \times 10^6 \; m^3 / s$ - has not changed. In the equatorial zone, integral transfer in the circulation cells of the Northern Hemisphere has slightly increased while that of the Southern Hemisphere has weakened. The smallest changes took place in the systems of the Southern Hemisphere currents. The redistribution of flow impulse took place everywhere mainly at the expense of strengthening of their stream character. The choice of turbulent parameters was varied in the course of calculations.

The boundary condition for ω at the contour G_2 was assumed to be the same as at the contour G_1, which led one to a strong change of the system of currents nearby the Antarctic Continent. Detailed examinations of the calculated variants are given in [18]. In the case of double-layered scheme of the ocean density, the calculated Gulf Stream flow rate has gone up by *30%* more, which allows us to expect further success in accounting for the synoptic eddy effect.

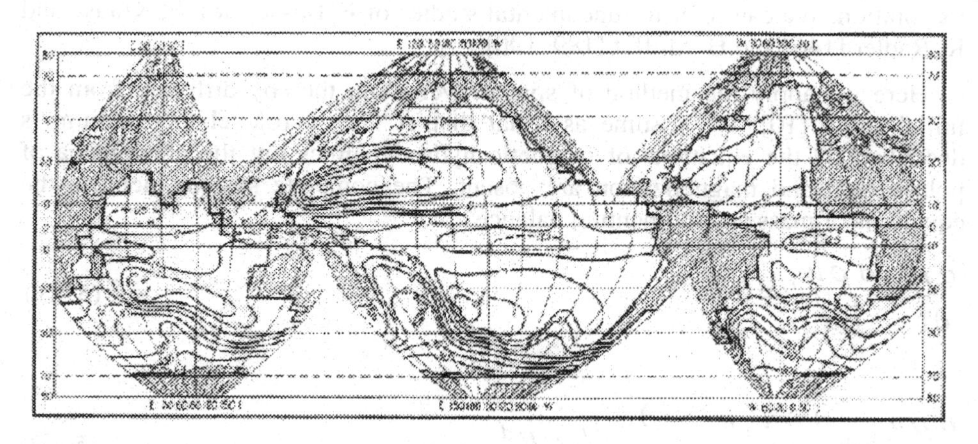

Figure 5.3.2. Stream function of World Ocean without account for synoptic Eddies

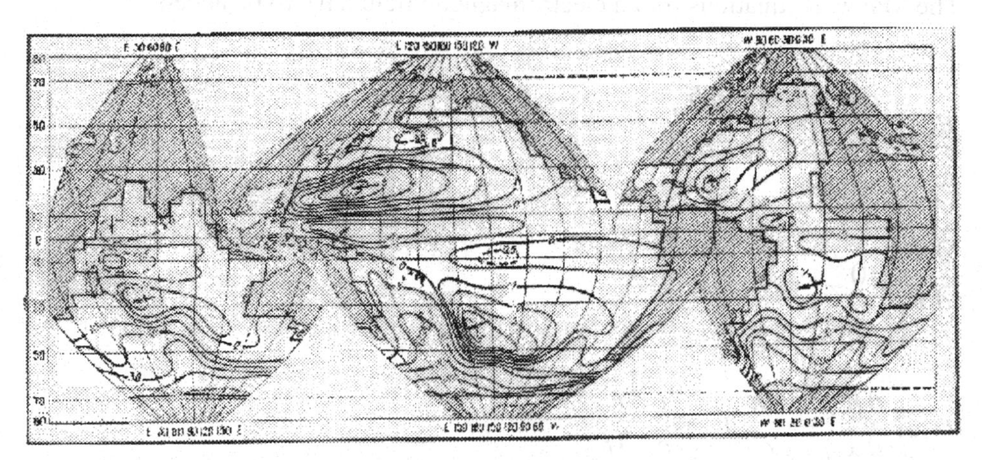

Figure 5.3.3. Stream function of World Ocean with account for synoptic eddies

5.4. TOWARDS THE PROBLEM OF A MAGNETIC DYNAMO[20]

Many authors have attacked the problem of the Earth Dynamo and the main conclusion was that the turbulence of electromagnetic viscous fluid is the key for its solution. We can refer to fundamental studies of F. Busse [38], F, Krause and K, Radler [146] and H. Moffat [189, 190].

Here we apply the method of spatial averaging, thereby differing from the initial study [10]. We assume as usual that at the microlevel the medium is described by the equations of magnetic hydrodynamics but the phenomena of polarization and magnetization are absent. The complete system includes the equations of mass and momentum balance

$$\frac{\partial \rho}{\partial t}+\frac{\partial \rho u_i}{\partial x_i}=0 \tag{5.4.1}$$

$$\frac{\partial \rho u_i}{\partial t}+\frac{\partial \rho u_i u_j}{\partial x_j}=\frac{\partial \sigma_{ij}}{\partial x_j}+F_i^l \tag{5.4.2}$$

The Maxwell equations for an electromagnetic field have to be added:

$$\varepsilon_{ijk}\frac{\partial e_k}{\partial x_j}=-\frac{1}{c}\frac{\partial h_i}{\partial t},$$

$$\frac{\partial h_i}{\partial x_i}=0 \tag{5.4.3}$$

$$\varepsilon_{ijk}\frac{\partial h_k}{\partial x_j}=\frac{4\pi}{c}j_i+\frac{1}{c}\frac{\partial e_i}{\partial t},$$

---

[20] This section is based on joint studies with O.Yu. Dinariev.

$$\frac{\partial e_i}{\partial x_i} = 4\pi \rho_e \qquad (5.4.4)$$

In the equations (5.4.1)-(5.4.4) the Gauss symmetric system of measurement units is used, e_i, h_i -intensity of electrical and magnetic fields, ρ_e-density of electric charge, $j_i = \rho_e u_i + j_{ci}$-total electric current, j_{ci}-conductivity current that is ruled by the Ohm law

$$j_{ci} = \sigma(e_i + \frac{1}{c}\varepsilon_{ijk} u_j h_k) \qquad (5.4.5)$$

σ - the conductivity coefficient and the Lorentz force is introduced:

$$F_{li} = \rho_e e_i + \frac{1}{c}\varepsilon_{ijk} j_j h_k$$

The condition of charge conservation is necessary for compatibility of the pair of equations (5.4.4).

$$\frac{\partial \rho_e}{\partial t} + \frac{\partial j_i}{\partial x_i} = 0 \qquad (5.4.6)$$

It is convenient to express the Lorentz force through components of energy - impulse tensor of the electromagnetic field

$$F_{li} = -\left(\frac{1}{c}\frac{\partial t_{i0}}{\partial t} + \frac{\partial t_{ij}}{\partial x_j}\right) \qquad (5.4.7)$$

$$t_{i0} = \frac{1}{4\pi}\varepsilon_{ijk} e_j h_k$$

$$t_{ij} = \frac{1}{4\pi}\left(-e_i e_j - h_i h_j + \frac{1}{2}(e_k e_k + h_k h_k)\delta_{ij}\right)$$

Besides equations (5.4.1)-(5.4.7), the equation of angular momentum has to be valid as a direct consequence of equations (5.4.4) and (5.4.7)

$$\frac{\partial}{\partial t}(\rho \varepsilon_{ijk} x_j u_k) + \frac{\partial}{\partial x_m}[\varepsilon_{ijk} x_j (\rho u_k u_m - \sigma_{km} + t_{km})] + \varepsilon_{ijk} x_j \frac{1}{c}\frac{\partial t_{k0}}{\partial t} = 0 \tag{5.4.8}$$

The equation for a magnetic field follows from the Maxwell equations (5.4.3) and (5.4.4) and the Ohm law (5.4.5)

$$\frac{\partial h_i}{\partial t} = \frac{\partial}{\partial x_j}\left(u_i h_j - u_j h_i\right) + \nu_m \frac{\partial^2 h_i}{\partial x_j \partial x_j} + \varepsilon_{ijk} \frac{\partial}{\partial x_j}\left(\frac{\partial e_k}{\partial t} + \frac{\partial e_l}{\partial x_l} u_k\right) \tag{5.4.9}$$

Here $\nu_m = c^2 / 4\pi\sigma$ is the magnetic viscosity.

If the changes of an electrical field and volumetric electrical charge are negligible, equation (5.4.9) will be reduced to the conventional equation for induction:

$$\frac{\partial h_i}{\partial t} = \frac{\partial}{\partial x_j}\left(u_i h_j - u_j h_i\right) + \nu_m \frac{\partial^2 h_i}{\partial x_j \partial x_j} \tag{5.4.10}$$

Averaging of equations (5.4.1), (5.4.2) and (5.4.8) taken into account, (5.4.7) results in the following equations for average variables

$$\frac{\partial \langle \rho \rangle}{\partial t} + \frac{\partial \langle \rho u_i \rangle_i}{\partial x_i} = 0 \tag{5.4.11}$$

$$\frac{\partial < \rho u_i >}{\partial t} + \frac{\partial < \rho u_i u_j >_j}{\partial x_j} = \frac{\partial < \sigma_{ij} >_j}{\partial x_j} -$$

$$\frac{\partial < t_{ij} >_j}{\partial x_j} - \frac{\partial < t_{i0} >}{c \partial t} \tag{5.4.12}$$

$$\varepsilon_{ijk} \frac{\partial}{\partial t} < \rho \xi_j u_k > + \varepsilon_{ijk} \frac{\partial}{\partial x_m} < \xi_j \times$$

$$(\rho u_k u_m - \sigma_{km} + t_{km}) >_m -$$

$$\varepsilon_{ijk} < \rho u_j u_k - \sigma_{jk} + t_{jk} >_k \tag{5.4.13}$$

$$+ \varepsilon_{ijk} \frac{\partial}{c \partial t} < \xi_j t_{k0} > = 0$$

For averaging of the Maxwell equations (5.4.3), (5.4.4) it is better to rewrite them in the integral form:

$$\oint_C e_i \, d l_i = - \frac{1}{c} \frac{\partial}{\partial t} \int_{S_C} h_i n_i \, d S,$$

$$\oint_C h_i \, d l_i = \frac{4 \pi}{c} \int_{S_C} j_i n_i \, d S + \frac{1}{c} \frac{\partial}{\partial t} \int_{S_C} e_i n_i \, d S, \tag{5.4.14}$$

$$\int_{\Delta S} h_i n_i \, d S = 0 \tag{5.4.15}$$

$$\int_{\Delta S} e_i n_i \, dS = 4\pi \int_{\Delta V} \rho_e \, dV$$

Here S_C is a two-dimensional surface bounded by a contour C, ΔS is a two-dimensional surface, limiting three-dimensional volume ΔV, n_i is a normal of the surface. In the case of a surface S_C the direction of normal n_i is supposed to be along the direction of detour of a contour C.

As one can see, the averaging of equations (5.4.14), including contour integrals, corresponds to *the averaging of a vector along a line.*

Considering boundary cross-sections ΔS_j of the elementary volume as the stretched surfaces S_C, over which the averaging of fluxes is being carried out, any closed sequences of edges ΔX_j of the volume ΔV will play the role of the contour C.

Therefore circulation of a vector in the left part of equations (5.4.14) breaks up into the sum of integrals of projections of a vector a_i at the edges taken along their consequence. It is reasonable to introduce the average vector $\{a_i\}_i$ of any vector a_i according to the formula

$$\{a_i\}_i = \frac{1}{\Delta X_i} \int_{\Delta X_i} a_i \, dx^i \tag{5.4.16}$$

Note that the remaining integral variables in equations (5.4.14) (5.4.15) do not demand additional definitions of spatial averaging.

We designate various averaged variables for mean electrical and magnetic fields as follows:

$$\{e_i\}_i = E_i, \qquad \{h_i\}_i = H_i$$

$$<e_i>_i = D_i, \quad <h_i>_i = B_i$$

According to the Stokes theorem, line averaging of a vector along a closed contour, consisting of edges of the elementary volume, determines in the total flux of vortexes through the corresponding cross-section.

The performance of this procedure in integrated equations (5.4.14), (5.4.15) allows us to pass to the macrodifferential equations, appropriate to the electromagnetic field with polarization and magnetization:

$$\varepsilon_{ijk}\frac{\partial E_k}{\partial X_j}=-\frac{1}{c}\frac{\partial B_i}{\partial t},$$

$$\varepsilon_{ijk}\frac{\partial H_k}{\partial X_j}=\frac{4\pi}{c}< j_i >_i+\frac{1}{c}\frac{\partial D_i}{\partial t} \tag{5.4.17}$$

$$\frac{\partial B_i}{\partial X_i}=0,$$

$$\frac{\partial D_i}{\partial X_i}=4\pi< \rho_e > \tag{5.4.18}$$

Thus, as it is easy to see, the averaging of equation (5.4.6) gives the balance equation for the averaged charge and current

$$\frac{\partial< \rho_e >}{\partial t}+\frac{\partial< j_i >_i}{\partial X_i}=0 \tag{5.4.19}$$

This provides compatibility of system (5.4.17) and (5.4.18).

Averaged equations (5.4.11) - (5.4.13), (5.4.17) and (5.4.18) for conducting media and electromagnetic fields contain more unknowns than the initial equations. Therefore for their completeness, it is necessary to invoke the constitutive laws, allowing us to reduce the number of unknown variables.

The form of equations (5.4.17), (5.4.19) allows us to assume that transition from one inertial system to another is accompanied by a standard linear transformation of pairs of vector E_i, B_i and D_i, H_i.

Therefore, it is sufficient to find the connection between these pairs of vectors in their own local coordinate system, that is, in the reference system, where at the space-time point

$$U_i = \langle \rho u_i \rangle / \langle \rho \rangle = 0 .$$

In the case of isotropic turbulence (but not necessary with mirror symmetry) the procedure of spatial averaging leads to a new isotropic medium, where the following relations are valid in linear approximation

$$D_i = \varepsilon_e E_i - \varepsilon_b B_i ,\ H_i = -\mu_e E_i + \mu_b^{-1} B_i \qquad (5.4.20)$$

Here ε_e and μ_b are analogs of dielectric and magnetic permeability coefficients. True dielectric and magnetic permeability coefficients are assumed to be exactly equal to unit. These $\varepsilon_e, \varepsilon_b, \mu_e, \mu_b$ appeared in (5.4.20) have a turbulent origin. In the special case of mirror symmetry we would have: $\varepsilon_h = \mu_e = 0$.

Let us consider the appearance of coefficients $\varepsilon_e, \varepsilon_b, \mu_e, \mu_b$ from the point of view of the microscopic theory. In Maxwell's equations (5.4.14), (5.4.15) there are two sources of an electromagnetic field: convective current $\rho_e u_i$ and conductivity current j_{ci}. The first one is determined by motion of external charges and is, in essence, a true source of an electromagnetic field, because in the absence of outer charges and motions the field will attenuate due to dissipation effects. The conductivity current is not equal to zero and can be a source of the field only in the case of some initial "seed" field, that could be created by outer charges.

Let us consider the case when there is some "seed" field:

$$e_i^0 = e_i^0 (t, x_j),\ h_i^0 = h_i^0 (t, x_j),$$

generating the initial conduction current in an isotropic turbulent medium (at $U_i = 0$). This current induces as additional electromagnetic field, which, in its turn, creates the conductivity current, etc. Note that, in general, a seed field inquires some induction charge ρ_e that depends linearly on the seed field to satisfy the law of charge conservation (5.4.6):

$$\rho_e = -\int_{-\infty}^{t} \frac{\partial j_k}{\partial x_k} dt = -\sigma \varepsilon_{kij} \int_{-\infty}^{t} \frac{\partial}{\partial x_k} (u_i h_j^0 (t, x_m)) dt =$$

$$= -\sigma \varepsilon_{kij} \int_{-\infty}^{t} \frac{\partial u_i}{\partial x_k} h_j^0 \, dt + \sigma c^{-1} u_i e_i^0$$

Correspondingly, the iteration solution of Maxwell's equations (5.4.3)-(5.4.5) leads to the integral expression

$$e_i (t, x_k) = \int (G_{ij}^{ee} (t, x_k; t_*, x_{k*}) e_j^0 (t_*, x_{k*}) +$$

$$+ G_{ij}^{eh} (t, x_k; t_*, x_{k*}) h_j^0 (t_*, x_{k*})) \, dt_* \, dx_{k*}$$

$$\tag{5.4.21}$$

$$h_i (t, x_k) = \int (G_{ij}^{he} (t, x_k; t_*, x_{k*}) e_j^0 (t_*, x_{k*}) +$$

$$+ G_{ij}^{hh} (t, x_k; t_*, x_{k*}) h_j^0 (t_*, x_{k*})) \, dt_* \, dx_{k*}$$

Here kernels $G_{ij}^{ee}, G_{ij}^{eh}, G_{ij}^{he}, G_{ij}^{hh}$ are random functions.

Let us assume that in the scale of ΔX_i any space averaging of core functions gives the delta – functions of time $\delta(t - t_*)$ and core values are negligible at space argument differences bigger than ΔX_i. Let us assume also that seed field $e_i^0 = e_i^0 (t, x_j)$, $h_i^0 = h_i^0 (t, x_j)$, is described by random functions, independent of kernels $G_{ij}^{ee}, G_{ij}^{eh}, G_{ij}^{he}, G_{ij}^{hh}$. Then spatial averaging will give local relations between seed and total fields:

$$E_i = g_{ij}^1 \langle e_j^0 \rangle + g_{ij}^2 \langle h_j^0 \rangle,$$

$$H_i = g^3_{ij} \langle e^0_j \rangle + g^4_{ij} \langle h^0_j \rangle$$

$$D_i = g^5_{ij} \langle e^0_j \rangle + g^6_{ij} \langle h^0_j \rangle,$$

$$B_i = g^7_{ij} \langle e^0_j \rangle + g^8_{ij} \langle h^0_j \rangle$$

$$g^1_{ij} = \{ \int G^{ee}_{ij}(t, x_k; t_*, x_{k*}) dt_* \, dx_{k*} \}_i$$

$$g^2_{ij} = \{ \int G^{eh}_{ij}(t, x_k; t_*, x_{k*}) dt_* \, dx_{k*} \}_i$$

$$g^3_{ij} = \{ \int G^{he}_{ij}(t, x_k; t_*, x_{k*}) dt_* \, dx_{k*} \}_i$$

$$g^4_{ij} = \{ \int G^{hh}_{ij}(t, x_k; t_*, x_{k*}) dt_* \, dx_{k*} \}_i$$

$$g^5_{ij} = \langle \int G^{ee}_{ij}(t, x_k; t_*, x_{k*}) dt_* \, dx_{k*} \rangle_i$$

$$g^6_{ij} = \langle \int G^{eh}_{ij}(t, x_k; t_*, x_{k*}) dt_* \, dx_{k*} \rangle_i$$

$$g^7_{ij} = \langle \int G^{he}_{ij}(t, x_k; t_*, x_{k*}) dt_* \, dx_{k*} \rangle_i$$

$$g^8_{ij} = \langle \int G^{hh}_{ij}(t, x_k; t_*, x_{k*}) dt_* \, dx_{k*} \rangle_i$$

Here, due to the medium isotropy, the matrices g^a_{ij} are proportional to unit matrices.

So, we obtain linear relations between vectors E_i, H_i, D_i, B_i, and $<e_j^0>,<h_j^0>$. Further exclusion of $<e_j^0>,<h_j^0>$ gives relations (5.4.20), where the coefficients $\varepsilon_e, \varepsilon_b, \mu_e, \mu_b$ are determined by statistical parameters of a turbulent medium.

Let us discuss an analogue of the Ohm law (5.4.5) for the averaged electrical current. We have

$$<j_{ci}>_i = <j_i>_i - <\rho_e u_i>_i = J_{ci} - <\rho_e' w_i>_i$$

Note that the average value microcurrent j_{ci} includes the conductivity current J_{ci} and turbulent pulsation transfer of charges. The total averaged electrical current can be stated as:

$$J_i = <\rho_e u_i>_i + \sigma(<e_i>_i + \frac{1}{c}\varepsilon_{ijk}<u_j h_k>_i) \tag{5.4.22}$$

At averaging (5.4.22) we shall use again the following expressions (at $i \neq j$):

$$<e_i>_j \equiv \{e_i\}_i \qquad <h_i>_j \equiv \{h_i\}_i$$

and $<u_i>_j \equiv \{u_i\}_i = U_i + v_i$. Here the circulation of v_i leads to a flux of χ_k through surface limited by the circulation contour (see Section 2.1).

In all cases we use pulsation components for electromagnetic variables:

$$\rho_e = <\rho_e> + \rho_e',$$

$$e_i = E_i + e_i' = D_i + e_i'',$$

$$h_i = B_i + h_i' = H_i + h_i''.$$

Then current (5.4.22) is represented as the sum of three terms:

$$J_i = J_i^{(1)} + J_i^{(2)} + J_i^{(3)}$$

$$J_i^{(1)} = <\rho_e> U_i$$

$$J_i^{(2)} = \sigma(D_i + \frac{1}{c}\varepsilon_{ijk} U_j H_k)$$

$$J_i^{(3)} = <\rho_e' w_i>_i + \frac{\sigma}{c}\varepsilon_{ijk} <w_j h_k''>_i$$

The current $J_i^{(1)}$ is caused by an average charge transport. It is necessary to interpret the current $J_i^{(2)}$ as a component of the conductivity current, caused by average fields. The current $J_i^{(3)}$ is a component of the conductivity current, caused by turbulent pulsations. In many cases the current $J_i^{(3)}$ can be neglected.

The expression derived for $J_i^{(2)}$, means proportionality of the conductivity current to a vector D_i in its own reference system. Taking into account for the constitutive expression (5.4.21), we deduce the generalized Ohm law for the averaged motion in its own reference system $J_i^{(2)} = \sigma_1 E_i - \sigma_2 B_i$, where $\sigma_1 = \sigma\varepsilon_e, \sigma_2 = \sigma\varepsilon_b$. Correspondingly, for any reference system it is necessary to use the expression

$$J_i^{(2)} = \sigma_1 (E_i + \frac{1}{c}\varepsilon_{ijk} U_j B_k) - \sigma_2 (B_i - \frac{1}{c}\varepsilon_{ijk} U_j E_k) \quad (5.4.23)$$

We apply now the averaging procedure at the energy - impulse tensor of the electromagnetic field. According to the developed method we obtain

$$<t_{i0}> = T_{i0} + T_{i0}'$$

$$<t_{ij}>_j = T_{ij}^{(1)} + T_{ij}^{(2)} + T_{ij}'$$

$$T_{i0} = \frac{1}{4\pi}\varepsilon_{ijk} D_j B_k,$$

$$T_{i0}' = \frac{1}{4\pi} \varepsilon_{ijk} e_j'' h_k''$$

$$T_{ij}^{(1)} = \frac{1}{4\pi}\left(-E_i D_j - B_i H_j + \frac{1}{2}(E_k E_k + H_k H_k)\delta_{ij} \right)$$

$$(T_{ij}^{(2)}) = diag\,(\Lambda_i),$$

$$\Lambda_i = \frac{1}{8\pi}((E_i - D_i)^2 + (B_i - H_i)^2)$$

$$T_{ij}' = \frac{1}{4\pi}\begin{pmatrix} -e_i' e_j'' - h_i'' h_j' + \\ \frac{1}{2}(e_k' e_k' + h_k' h_k')\delta_{ij} \end{pmatrix} + diag(\Lambda_i')$$

$$\Lambda_i' = \frac{1}{8\pi}((e_i' - e_i'')^2 + (h_i' - h_i'')^2)$$

Averaging of dynamics equations of the medium (5.4.11) - (5.4.13) gives the following equations for average fields:

$$\frac{\partial \rho}{\partial t} + \frac{\partial \rho U_i}{\partial X_i} = 0 \tag{5.4.24}$$

$$\frac{\partial}{\partial t}(\rho U_i + \frac{1}{c} T_{i0}) + \frac{\partial \rho U_i U_j}{\partial X_j} = \\ -\frac{\partial p}{\partial X_i} + \frac{\partial R_{ij}^m}{\partial X_j} - \frac{\partial (T_{ij}^{(1)} + T_{ij}^{(2)})}{\partial X_j} - \frac{\partial T_{i0}'}{c\,\partial t} \tag{5.4.25}$$

$$\frac{\partial}{\partial t}(\rho M_i + \Im_i) + \frac{\partial \rho M_i U_j}{\partial X_j} = \tag{5.4.26}$$

$$\frac{\partial \mu_{ij}}{\partial X_j} + \frac{\partial \mu^e_{ij}}{\partial X_j} - \varepsilon_{ijk} R^m_{jk} + \varepsilon_{ijk} T^{(1)}_{jk}$$

Here $R^m_{ij} = (R_{ij} - T_{ij}')$ is the modified Reynolds stress tensor, which is, in general, asymmetric,

$$\Im_i = \frac{1}{c} \varepsilon_{ijk} < \xi_j t_{k0} >$$

is the angular momentum of an electromagnetic field, $\mu^e_{ij} = -\varepsilon_{ikl} < \xi_k t_{lj} >_j$ is a couple stress tensor of the electromagnetic field.

As the number of unknown variables in the set of equations (5.4.19) - (5.4.21), (5.4.23)-(5.4.26) still exceeds the number of equations, some additional hypotheses are necessary for completeness.

Let us accept that pulsation stresses, having an electromagnetic origin, are not essential in the expression for the Reynolds modified tensor. We neglect also T_{i0}'. Therefore, as earlier, we accept

$$R^m_{ij} = R_{ij} = 2\nu\rho\left(\frac{\partial U_i}{\partial X_j} + \frac{\partial U_j}{\partial X_i}\right) + 2\gamma\rho\,\varepsilon_{ijk}\omega_k$$

$$\mu_{ij} = 2\eta\rho \frac{\partial(\Omega_i + \omega_i)}{\partial X_j}$$

For the moment characteristics of the electromagnetic field we shall accept the constitutive laws on the basis of the correspondence of transformation properties of interconnected variables

$$\Im_i = \zeta\left(B_i - \frac{1}{c}\varepsilon_{ijk} U_j E_k\right) + \xi < \rho_e > \omega_i\ ,$$

$$\mu_{ij}^{e} = \nu_e \frac{\partial \mu_i}{\partial X_j}$$

These last expressions complete the magneto-hydrodynamic model. The model describes a wide class of the phenomena, in particular, the magnetic dynamo effect.

Let us recall that according to the modern point of view the generation of the Earth's magnetic field by hydrodynamic currents is connected with the presence of the dependence on a magnetic field in the Ohm law, see (5.4.23).

Particularly, let us consider a medium, in which there is no mean density of a charge, and the variables:

$$J_i^{(3)}, \qquad \frac{1}{c}\frac{\partial D_i}{\partial t}, \qquad \frac{\sigma_2}{c\sigma_1} U_i$$

are negligibly small. Then equations (5.4.19), (5.4.20), (5.4.23) yield the following equation of the averaged magnetic field induction:

$$\frac{\partial B_i}{\partial t} = \frac{\partial}{\partial x_j}\left(U_i B_j - U_j B_i\right) + \nu_m' \frac{\partial^2 B_i}{\partial x_j \partial x_j} + \alpha \varepsilon_{ijk} \frac{\partial B_k}{\partial x_j} \tag{5.4.27}$$

This equation differs from equation (5.4.10) by the last term on the right side that is known as the α-term where:

$$\alpha = \frac{c\sigma_2}{\sigma_1}.$$

This term [146, 253] provides the existence of exponentially growing modes of the magnetic field.

Note that the magnetic viscosity

$$\nu_m' = \frac{c^2}{4\pi\sigma_1}$$

in macroequation (5.4.27) differs from magnetic viscosity ν_m in microequation (5.4.10). This distinction is determined by the factor ε_e, reflecting the difference of averaging result of a microscopic electrical field along a line and over the surface.

Under the non-induction approximation, the equation system of the MHD turbulence obtained for incompressible fluid under the stationary magnetic field has the form:

$$\frac{\partial U_i}{\partial X_i} = 0, \tag{5.4.28}$$

$$\rho\left(\frac{\partial U_i}{\partial t} + \frac{\partial U_i U_j}{\partial X_j}\right) = -\frac{\partial p}{\partial X_i} + \frac{\partial R_{ij}}{\partial X_j} + \frac{\sigma}{c}\varepsilon_{ijk} B_k \left(E_j + \frac{1}{c}\varepsilon_{jmn} U_m B_n\right), \tag{5.4.29}$$

$$\rho\left(\frac{\partial M_i}{\partial t} + \frac{\partial M_i U_j}{\partial X_j}\right) = \frac{\partial \mu_{ij}}{\partial X_j} + \varepsilon_{ijk} R_{jk} + \varepsilon_{ijk} M_j B_k, \tag{5.4.30}$$

$$E_i = -\frac{\partial \varphi}{\partial X_i}, \qquad \frac{\partial^2 \varphi}{\partial X_i \partial X_i} = \frac{1}{c}\frac{\partial}{\partial X_i}(\varepsilon_{ijk} U_j B_k). \tag{5.4.31}$$

In this case, the action of the magnetic field on the flow of the conductive fluid occurs due to a force

$$F_i = \frac{\sigma}{c} \varepsilon_{ijk} B_k \left(E_j + \frac{1}{c} \varepsilon_{jmn} U_m B_n \right), \tag{5.4.32}$$

This changes the distribution of the velocity field and of the moment $C_i = \varepsilon_{ijk} M_j B_k$, conspiring to combine the rotational axis of the turbulent eddy with the direction of the magnetic field. Thereby there is a definite analogy between the turbulent flow of the conductive fluid and the flow of the ferromagnetic fluid inside the external magnetic field that was noted in [301] in order to explain the effect of three-dimensional turbulence reconstruction into a two-dimensional one.

This idea corresponds to a variant of the theory noted in [10], where the moment $(\sigma / 2c) \varepsilon_{ijk} M_j B_k$ was introduced into the magnetic moment of the turbulent eddy. The qualitative calculations of three-dimensional turbulence dying down are presented in [106, 200-203] within the scope of a model analogous with (5.4.28)-(5.4.32).

Let us note that I. Tamm emphasized (see his known course on electricity [274] of 1949) that the Maxwell stress tensor could be non-symmetrical in macroscale, though the microstress tensor is symmetrical. He adduced the effect of polarization of elastic dielectrics as the example. Aspects of the Cosserat theory application to the electromagnetic continua were discussed by C. Truesdell [289], in lectures by R. Stojanovic [269]), by G. Maugin and A. Eringen [183, 184], et al as well as in monographs on electrodynamics (see [64]).

CHAPTER 6

VORTEX SYMMETRY IN STATISTICAL THEORY

6.1. HOMOGENEITY AND SYMMETRY CONCEPTS

Statistical theory of turbulence was initiated by G.I. Taylor [279] who proposed a characterization of a turbulence field by moment functions, that is, by mean values of products of field variables at neighboring points. In practice it corresponded to measurements of velocity changes in time by thermo - anemometers and by further averaging in time.

Although we know that a proper continuum and engineering calculations of turbulence dynamics needs to be averaged over space (over volume, surfaces and even along lines of coordinates or grid cells), in both cases the hypotheses that the total ensemble of random realizations contains in time or in space is accepted correspondingly. Therefore the dependence of stress tensors on spin velocity has to be found in the measured statistical moments of random turbulence fields if one assumes the existence of angular symmetry, as well as the axial symmetry studied by G. Batchelor [19] and S. Chandrasekhar [40, 41].

Moreover, very often in current calculation practice the statistical averaged parameters are used jointly with the Reynolds equations developed in the continuum approach, realized by space averaging. Of course, in these cases an adequate symmetry has to be assured.

In statistical theory the averaging is performed over the random parameter of the realization ensemble, assuming that it is equivalent to averaging in time. We shall denote such averaging by an upper bar. For example, the two-point moments of local pulsation velocities (invariant to rigid translations) has the following form:

$$Q_{ij} = \overline{w_i(x_1)\, w_j(x_2)} \tag{6.1.1}$$

The turbulent random field is *homogeneous* [20, 195] if there is no dependency of an averaged two-point function F^* on a rigid translation of the measurement points, that is:

$$\overline{F^*(x_i - y_i; x_i' - y_i)} = \overline{F^*(x_i - x_i')} \equiv \overline{F^*(r_i)} \equiv F(r_i) \tag{6.1.2}$$

Here y_i is an arbitrary translation vector, $r_i = x_i - x'_i$ is a radius-vector, connecting the points of measurements.

Conventional homogeneity assumes that average velocity U = *const,* as it happens, for example, behind a grid in an aerodynamic tube. For simplicity, the analysis is performed usually in a system of coordinates where U = *0.* Then (6.1.1) includes just velocity pulsation values and this object has to be invariant under rigid translations.

According to Batchelor [20], §4, the important consequence of homogeneity follows for the velocity correlation (6.1.1), that is

$$Q_{ij}(r_k) = Q_{ji}(-r_k) \tag{6.1.3}$$

In reality, the Batchelor condition is not a consequence of (6.1.1) and, as we shall see further, would exclude any asymmetrical objects from consideration without symmetry requirements. Equality (6.1.3) is based implicitly also on the symmetry property of mirror reflections. Only in this case the rigid translation at a distance r can be described with a vector $\boldsymbol{r}$.

To show this, let us compare its left-hand and right-hand sides. At the left-hand side we have:

$$Q_{ij}(r_k) = \overline{w_i(x_k) w_j(x_k + r_k)} \tag{6.1.4}$$

If to add y_k = - r_k to both arguments at the right-hand side of (6.1.4), then we obtain

$$Q_{ij}(r_k + y_k) = \overline{w_i(x_k - r_k) w_j(x_k)} \tag{6.1.5}$$

According to the homogeneity concept, the left-hand side would be the same as in (6.1.4), although the right-hand side was changed. But it is not so: the addition of y_k = - r_k contains simultaneous reorientation of vector argument r_k to the opposite as it is shown in Figure 6.1.1, where the proper vector components are attached to the radius-vector. Here variant **A** corresponds to (6.1.4) and **B** to (6.1.5). They are different because a third radius-vector is involved besides two velocity pulsation vectors.

Equality (6.1.5) can be transformed indeed to the right-hand side of (6.1.3):

$$\overline{w_i(x_k - r_k)w_j(x_k)} \equiv \overline{w_j(x_k)w_i(x_k - r_k)} \equiv Q_{ji}(-r_k) \tag{6.1.6}$$

However, the left sides of (6.1.4) and (6.1.5) are not equivalent and (6.1.3) does not follow only from the homogeneity concept.

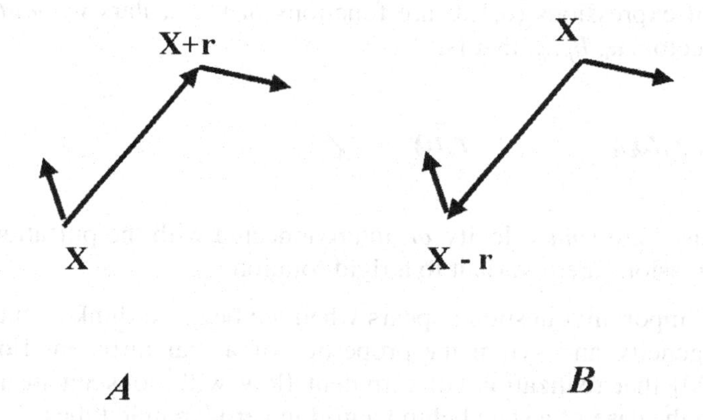

Figure 6.1. Two-point velocity correlation tensor includes an oriented radius vector

Let us consider symmetry properties of a random turbulent field. They can be accounted for if we apply Robertson's [248] ideas, developed in [20]. Accordingly, scalar products

$$f = F_{ij}\ldots a_i b_j \ldots \tag{6.1.7}$$

of two-point moments $F_{ij.....}$ and arbitrary vectors a_i, b_j... have to be invariant under any transformations that assume turbulent field symmetry. All these expressions have to be scalar functions of products of *vector* ***r***, connecting the selected points, of *spin tensor* $\omega_{ij} = \varepsilon_{ijk}\omega_k$, (invariant to rigid rotation) as well as *arbitrary vectors* a_i, b_j.... In other words [36, 37],

$$f = f(r^2, r_i a_i, a_j b_j, \omega_k a_k, r_i \omega_i, \omega_k \omega_k, \varepsilon_{ijk} a_i r_j \omega_k, ...) \quad (6.1.8)$$

Because f is a *linear* function of arbitrary vector a_i and b_j...by definition (6.1.7), we have

$$F_{ij} = A_0 + A_1 r_i r_j + A_2 \omega_i \omega_j + A_3 r_i \omega_j \quad (6.1.9)$$

Here pseudo-scalars as $\varepsilon_{ijk} a_j b_j r_k$, $\varepsilon_{ijk} a_j b_j \omega_k$ will not be considered and $\omega^2 = 1$. The coefficients of expressions (6.1.8) are functions of the scalars *without inclusion* of arbitrary vectors a_i, b_j,..., that is,

$$A_m = A_m(r^2, r_i \omega_i), \qquad r_i \omega_i = r\chi \quad (6.1.10)$$

Because we use here *spin* velocity ω_i, interconnected with the pulsation field, all resulting expressions are invariant to a rigid rotation.

One more important question appears when we begin to think about adequacy of the homogeneity and symmetry properties of a real turbulent flow. It was mentioned [20] that realizations of turbulent flow with constant mean velocity correspond to the case of a state behind a grid in aerodynamic tubes.

In more common turbulent flows the mean velocity changes in space and its profile is linear at a scale ΔX of spatial averaging (Figure 6.1.2), see also [93]. The selection of a smaller space scale, that could be characterized by practically constant average velocity, contradicts the existence of turbulent eddies with scale of λ order.

However, this random velocity field can be considered also as *homogeneous* if the *distortion average rate is constant*, that is,

$$\overline{\frac{\partial u_i}{\partial x_j}} \equiv \frac{\partial U_i}{\partial X_j} = const \neq 0 . \quad (6.1.11)$$

This situation can be called as *gradiental homogeneous,* see also [93] and exactly corresponds to the Kolmogorov "local homogeneity" [138-140]. The structure function $\boldsymbol{D_{ij}}$, selected by Kolmogorov for statistical description of such cases, includes differences of variables at close points:

$$\Delta_i(x_k, r_j) = u_i(x_k) - u_i(x_k + r_k) \tag{6.1.12}$$

This object is adequate for the statistical study of turbulent eddies because it is invariant to rigid translations of coordinate system, see (6.1.2), and at short distances $\boldsymbol{r_k}$ (6.3.2) includes values, averaged in (6.1.11). In fact, its study focuses on a random vorticity (a skew part of the distortion tensor) or/and strain rate (its symmetrical part):

$$D_{ij} = \overline{\Delta_i(r_m)\Delta_j(r_k)} \approx \overline{\frac{\partial u_i}{\partial x_m}\frac{\partial u_j}{\partial x_k}} r_m r_k \tag{6.1.13}$$

This corresponds to a distortion random field.

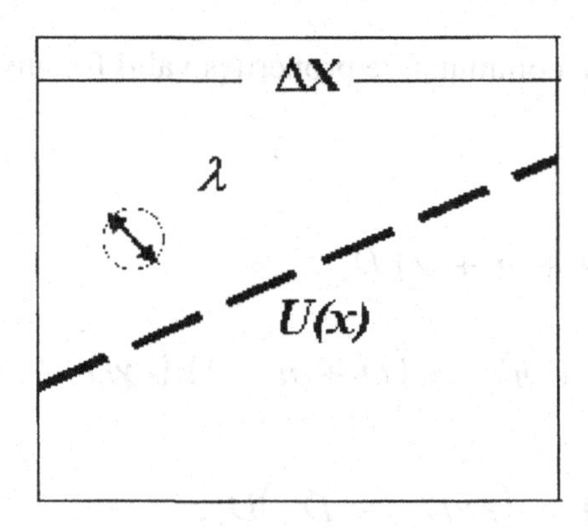

Figure 6.1.2. Scales of spatial averaging and turbulent eddy in the average velocity field

It is reasonable to think that the idea of Kolmogorov to connect statistically mean values of D_{ij} in a homogeneous field with turbulent energy flux is in accordance with the link Q_{ij} of a homogeneous field with turbulent energy itself.

6.2. TENSORIAL MOMENTS OF A VELOCITY PULSATION FIELD

Moments of the first order. In accordance with the previous results, the correlation vector function has the following form (term $l_3\omega_i$ is omitted):

$$L_i = l_1 r_i + l_2 \varepsilon_{ijk}\omega_k r_j \tag{6.2.1}$$

The turbulent field is solenoidal[21] and therefore (6.2.1) can be simplified. For this purpose, the following operators [37, 40, 41] will be used:

$$\nabla_i = \frac{\partial}{\partial r_i} \equiv r_i D_r + \omega_i D_\chi;$$
$$D_r = \frac{1}{r}\frac{\partial}{\partial r} - \frac{\chi}{r^2}\frac{\partial}{\partial \chi}; \qquad D_\chi = \frac{1}{\chi}\frac{\partial}{\partial \chi} \tag{6.2.2}$$

Here we shall use the commutative properties valid for any n:

$$D_r D_\chi = D_\chi D_r,$$
$$D_r (D + n) = (D + n + 2) D_r \tag{6.2.3}$$

$$\left(r\chi D_r + D_\chi\right)(D + n) = (D + n + 1)\left(r\chi D_r + D_\chi\right)$$

$$D_r\left(r\chi D_r + D_\chi\right) = \left(r\chi D_r + D_\chi\right)D_r,$$
$$D = r(\partial / \partial r), \qquad D_r f(r\chi) = 0 \tag{6.2.4}$$

---

[21] An arbitrary vector field is solenoidal if the divergence of the field vector equals zero.

The solenoidal condition for L_i means:

$$\nabla_i L_i = (D + 3) l_1 = 0 \tag{6.2.5}$$

This differential equation will give $l_1 \equiv 0$ because l_1 is limited at infinity and at $r \to 0$. Therefore, a limited solenoidal vector will be in general as

$$L_i = L \varepsilon_{ijk} \omega_j r_k, \qquad L = l_2(r^2, r\chi, t) \tag{6.2.6}$$

The gradient of vector L_i is:

$$\nabla_i L_j = -\varepsilon_{ijk} \omega_k L + r_i \varepsilon_{kmn} \omega_m r_n D_r L + \omega_i \varepsilon_{kmn} \omega_m r_n D_\chi L \tag{6.2.7}$$

Moments of the second order. The most important is the correlation function constructed in accordance with (6.1.7):

$$Q_{ij} = q_0 \delta_{ij} + q_1 r_i r_j + q_2 \varepsilon_{ijk} \omega_k + q_3 \varepsilon_{imk} \omega_k r_m r_j +$$
$$q_4 \varepsilon_{jmk} \omega_k r_m r_i + q_5 \varepsilon_{imk} \omega_k r_m \omega_j + q_6 \varepsilon_{jmk} \omega_k r_m \omega_i \tag{6.2.8}$$

It may include symmetrical and skew terms. Because the scalar $Q_{ij}\, a_i\, b_j$ is invariant to any transformations,

$$q_k(r) = q_k(-r), \quad k = 0,\dots,4; \qquad q_k(r) = -q_k(-r), \quad k = 5, 6.$$

Tensor Q_{ij} is assumed to be solenoidal by index j. Therefore:

$$(D + 4)\, q_0 + D_r q_1 = 0;$$

$$D_r q_2 + (D + 4) q_3 + q_4 + (r\chi D_r + D_\chi)\, q_5 = 0 \tag{6.2.9}$$

$$D_\chi q_1 = 0$$

Analogously, the solenoidal second index of Q_{ij}, that is, $\partial Q_{ij} / \partial x_j = 0$, means:

$$(D + 4) q_0 + D_r q_1 = 0; \qquad D_\chi q_1 = 0$$

$$q_3 - D_r q_2 + (D + 4) q_4 + (r\chi D_r + D_\chi) q_6 = 0 \tag{6.2.10}$$

If tensor Q_{ij} is solenoidal by both indexes, let us consider the coefficients:

$$M_s = \frac{1}{2}(q_3 + q_4);\ M_a = \frac{1}{2}(q_3 - q_4);$$

$$N_s = \frac{1}{2}(q_5 + q_6);\ N_a = \frac{1}{2}(q_5 - q_6). \tag{6.2.11}$$

The following limitations are valid for them because of the solenoidal conditions:

$$(D + 5) M_s + (r\chi D_r + D_\chi) N_s = 0$$

$$D_r q_2 + (D + 3) M_a + (r\chi D_r + D_\chi) N_a = 0 \tag{6.2.12}$$

Using the commutative relations (6.2.3), (6.2.4) let us determine all variables q_i, N_i, M_i in such a manner that all solenoidal limitations will be satisfied identically. This means that the scalar potentials Q_k are introduced in a such way that:

$$Q_{ij} = -\frac{1}{r} r_i r_j \frac{\partial Q_0}{\partial r} + \delta_{ij}(D + 2)Q_0 + \omega_{ij}[(D + 1)Q_2 +$$

$$(r\chi D_r + D_\chi)Q_3] - (\omega_{im} r_j - \omega_{jm} r_i) r_m D_r Q_2 - (\omega_{im}\omega_j - \qquad (6.2.13)$$

$$\omega_{jm}\omega_i) r_m D_r Q_3 - (\omega_{im} r_j + \omega_{jm} r_i) r_m (r\chi D_r + D_\chi) Q_1 +$$

$$(\omega_{im}\omega_j + \omega_{jm}\omega_i)(D + 4)Q_1$$

Potentials Q_k satisfy the evenness conditions that are a consequence of the same ones for values q_k :

$$Q_r(r) = Q_r(-r), \qquad k = 0,\ 2;$$

$$(6.2.14)$$

$$Q_r(r) = -Q_r(-r), \qquad k = 1,\ 3$$

Moments of the third order. Again using (6.2.3), (6.2.4), we can represent the correlation tensor Π_{ijk}, symmetric in indices i and j, as:

$$\Pi_{ijk} = \pi_1 r_i r_j r_k + \pi_2(r_i\delta_{jk} + r_j\delta_{ik}) + \pi_2 r_k\delta_{ij} + \pi_4(r_i\omega_{jk} + r_j\omega_{ik}) +$$

$$\pi_5(\omega_{im}\delta_{jk} + \omega_{jm}\delta_{ik}) + \pi_6 r_m\omega_{km}\delta_{ij} + \pi_7(r_i\omega_{jm} + r_j\omega_{im}) r_m r_k + \qquad (6.2.15)$$

$$\pi_8 r_m\omega_{km} r_i r_j + \pi_9(r_i\omega_{jm} + r_j\omega_{im})\omega_{kn} r_m r_n + \pi_{10}\omega_{jn}\omega_{im} r_m r_k r_n +$$

$$\pi_{11}(\omega_{im}\omega_{jk} + \omega_{jm}\omega_{ik}) r_m + \pi_{12}\omega_{jn}\omega_{im}\omega_{kl} r_m r_n r_l + \pi_{13}(\omega_i\omega_{jk} +$$

$$\omega_j\omega_{ik}) + \pi_{14}(r_i\omega_{jm} + r_j\omega_{im}) r_m r_k + \pi_{15}(\omega_i\omega_{jm} + \omega_j\omega_{im})\omega_k r_m +$$

$$\pi_{16}(r_i\omega_{jm} + r_j\omega_{im})r_m\omega_k + \pi_{15}(\omega_i\omega_{jm} + \omega_j\omega_{im})\omega_k r_m + \pi_{18}\omega_{km}r_m \times$$

$$(\omega_i r_j + \omega_j r_i) + \pi_{19}(\omega_i\omega_{jm} + \omega_j\omega_{im})\omega_{kn}r_m r_n + \pi_{20}\omega_{im}\omega_{jn}\omega_k r_m r_n$$

If we assume that $\Pi_{ijk}\, a_i\, b_j\, c_k$ is a true scalar, the evenness conditions for the coefficients π_k will follow.

The solenoidal conditions for Π_{ijk} over the last index lead to the equations:

$$(D + 5)\pi_1 + 2D_r\pi_2 = 0;$$

$$2\pi_2 + (D + 3)\pi_3 = 0;$$

$$D_\chi\pi_2 = 0; \tag{6.2.16}$$

$$D_r(\pi_4 + \pi_5) + (D + 5)\pi_7 + \pi_8 + (r\chi D_r + D_\chi)\pi_{16} = 0;$$

$$D_\chi\pi_5 + D_r\pi_{13} + (D + 4)\pi_{14} +$$

$$(r\chi D_r + D_\chi)\pi_{15} + \pi_{16} + \pi_{18} = 0;$$

$$(D + 5)\pi_{10} + 2D_r\pi_{11} + (r\chi D_r + D_\chi)\pi_{20} = 0,$$

$$\pi_9 = \pi_{11} = \pi_{12} = \pi_{19} = 0$$

These equations are satisfied identically if one defines the values π_k through scalar potentials Π_k :

$$\pi_1 = \frac{2}{r}\frac{\partial}{\partial r}\Pi_1; \qquad \pi_3 = 2\Pi_1;$$

$$\pi_2 = -(D+3)\Pi_1 \equiv -(r\frac{\partial}{\partial r} + 3)\Pi_1;$$

$$\pi_4 = \Pi_2; \qquad \pi_5 = \Pi_3; \qquad \pi_6 = \Pi_1;$$

$$\pi_7 = \Pi_5; \qquad \pi_{13} = \Pi_6 \qquad \pi_{14} = \Pi_7$$

$$\pi_{15} = \Pi_8; \qquad \pi_{16} = \Pi_9; \qquad \pi_{17} = \Pi_{10};$$

$$\pi_8 = -D_r(\Pi_2 + \Pi_8) - (D+5)\Pi_5 - (r\chi D_r + D_\chi)\Pi_9$$

$$\pi_{18} = -D_\chi\Pi_3 - D_r\Pi_6 - (D+4)\Pi_7 - (r\chi D_r + D_\chi)\Pi_8 - \Pi_9;$$

$$\pi_{19} = -(r\chi D_r + D_\chi)\Pi_{11}; \qquad \pi_{20} = (D+5)\Pi_{11}.$$

Using the previous potential expressions, the tensor Π_{ijk} with solenoidal index k:

$$\Pi_{ijk} = r_i r_j r_k \frac{2}{r}\frac{\partial}{\partial r}\Pi_1 - (r_i\delta_{jk} + r_j\delta_{ik})(D+3)\Pi_1 + +2r_k\delta_{ij}\Pi_1 +$$

$$(r_i\omega_{jk} + r_j\omega_{ik})\Pi_2 + (\omega_{im}\delta_{jk} + \omega_{jm}\delta_{ik})r_m\Pi_3 + \omega_{km}\delta_{ij}r_m\Pi_1 +$$

$$(r_i\omega_{jk} + r_j\omega_{ik})\Pi_2 + (\omega_{im}\delta_{jk} + \omega_{jm}\delta_{ik})r_m\Pi_3 + \omega_{km}\delta_{ij}r_m\Pi_1 + \tag{6.2.17}$$

$$((r_i\omega_{jm} + r_j\omega_{im})r_m r_k\Pi_5 - \omega_{km}r_i r_j r_m[D_r(\Pi_2 + \Pi_3) + (D+5)\Pi_3 +$$

$$(r\chi D_r + D_\chi)\Pi_9] - \omega_{im}\omega_{jn}r_k r_n r_m(r\chi D_r + D_\chi)\Pi_{11} + \omega_{km}\omega_j\omega_i r_m\Pi_{10} +$$

$$(\omega_i\omega_{jk} + \omega_j\omega_{ik})\Pi_6 + (\omega_i\omega_{jm} + \omega_j\omega_{im})r_k(r_m\Pi_7 + \omega_m\Pi_8) +$$

$$(r_i\omega_{jm} + r_j\omega_{im})r_m\omega_k\Pi_9 - \omega_{km}r_m(r_i\omega_j + r_j\omega_i)[D_\chi\Pi_3 + D_r\Pi_6 +$$

$$+ (D+4)\Pi_7 + (r\chi D_r + D_\chi)\Pi_8 + \Pi_9] + \omega_{im}\omega_{jn}r_k r_n r_m(D+4)\Pi_{11}.$$

Here $\Pi_k = \Pi_k(r^2, r\chi, t)$, $k > 1$. We can get also the evenness conditions for eleven potentials in (6.2.15).

In the isotropic case only terms with Π_1 are non-zero. All others reflect the vortex anisotropy.

6.3. TURBULENCE WITH REFLECTION SYMMETRY

Let consider the following tensors:

$$L_i = \overline{pw_i^{\prime}}, \quad Q_{ij} = \overline{w_i w_j^{\prime}}, \quad \Pi_{ijk} = \overline{w_i w_j w_k^{\prime}} \tag{6.3.1}$$

Here p, $p^{\prime}$ and $\mathbf{w}$, $\mathbf{w}^{\prime}$ are pressure and pulsation velocity values at points M and $M^{\prime}$, correspondingly, $\mathbf{r} = \mathbf{x}_{M^{\prime}} - \mathbf{x}_M$ and "bar" means the averaging over the ensemble of turbulent field realizations.

Under the assumption of commutability of differentiation and the ensemble averaging, the Navier – Stokes equations for incompressible fluids will give us the following equalities:

$$\nabla_i L_i = \nabla_i Q_{ij} = \nabla_j Q_{ij} = \nabla_k \Pi_{ijk} = 0 \tag{6.3.2}$$

$$\frac{\partial Q_{ij}}{\partial t} = 2\nu \nabla_k \nabla_k Q_{ij} + S_{ij} \tag{6.3.3}$$

$$S_{ij} = \nabla_i (\Pi_{ikj} - \Pi^*_{ikj}) + \frac{1}{\rho}(\nabla_i L_j - \nabla_j L^*_i)$$

Here ρ and ν are fluid density and kinematics' viscosity as usual and

$$\Pi^*_{ijk} = \overline{w_i w'_j w'_k}; \quad L^*_i = \overline{p' w_i} \tag{6.3.4}$$

In accordance with solenoid conditions (6.3.2) tensors (6.3.1) can be represented as (6.2.6), (6.2.13) and (6.2.17) correspondingly.

Tensors S_{ij} and $\Gamma_{ij} = \nabla_k \nabla_k Q_{ij}$, as it can be shown, are also solenoidal for both indexes and can be represented through their potentials S_k and Γ_k as in (6.2.13).

The conditions of turbulence homogeneity additionally limit the introduced scalar potentials. So, $L(r) = -L^*(-r)$, where L^* plays the same role for L^*_i as L for L_i.

Let us note that the symmetrical part of (6.2.13) satisfies the Batchelor condition (6.1.3) identically but the asymmetric part has to be zero:

$$Q_3 = 0, \qquad Q_4 = 0 \tag{6.3.5}$$

In this special case of mirror symmetry, equation (6.2.3) can be transformed if we calculate the potentials Γ_k:

$$\Gamma_1 = \frac{1}{r^4}\frac{\partial}{\partial r}(r^4 \frac{\partial}{\partial r})Q_1;$$

$$\Gamma_2 = \Delta_6 Q_2,$$

$$\Delta_n = r^2 D_r^2 + 2r\chi D_r D_\chi + D_\chi^2 + n + 1 \equiv$$

$$\frac{1}{r^n}\frac{\partial}{\partial r}(r^n \frac{\partial}{\partial r}) + \frac{1-\chi^2}{r^2}\frac{\partial^2}{\partial \chi^2} - \frac{n\chi}{r^2}\frac{\partial}{\partial \chi}$$

Now we can get from (6.3.2) the following equations:

$$\frac{\partial Q_1}{\partial t} = \nu\Delta_4 Q_1 + S_1, \qquad S_3 = 0;$$
$$\frac{\partial Q_2}{\partial t} = \nu\Delta_6 Q_2 + S_2, \qquad S_4 = 0 \tag{6.3.6}$$

The first equation of (6.3.6) is the famous Karman-Howarth equation [130] for isotropic turbulence. Potentials S_k correspond to inertial terms.

Consider the terms in Q_{ij}, depending on vortex anisotropy even under (6.1.3). We use now a coordinate system with the origin at point M and axis x_1 coinciding with spin velocity $\boldsymbol{\omega}$ Vector $\boldsymbol{r}$ is in the plane x_1, x_2.

Let the component $\omega_1 = 1$. Then the Q_{ij} components can be expressed as:

$$Q_{11} = Q_{22} = q_1 r_1^2 + q_2;$$
$$Q_{33} = q_2;$$
$$Q_{12} = Q_{21} = q_1 r_1 r_2; \tag{6.3.7}$$
$$Q_{13} = Q_{31} = -M_s r_1 r_2 - N_s r_2;$$
$$Q_{23} = Q_{32} = -M_s r_2^2.$$

Here Q_{13}, Q_{31}, Q_{23} and Q_{32} are generated by the vortex anisotropy. The scalars appearing here were introduced earlier.

The energy of turbulent pulsations will be the same as in the case of isotropy:

$$Q_{11} + Q_{22} + Q_{33} = q_1 r^2 + 3q_2. \tag{6.3.8}$$

Calculate now the longitudinal $Q_{//}$ and lateral $Q_{\perp}$ correlations of velocities. For this purpose, axis x_1 will be along vector $\boldsymbol{r}$ and axis x_2 – along a lateral velocity component at point M. Then we can see the additional term, which is evidently connected with anisotropy properties:

$$Q_{//} = q_1 r^2 + q_2$$
$$Q_{\perp} = q_2 - 2N_5 r\omega_2\omega_3 = q_2 \mp 2N_5 r\beta[(1-\beta^2)(1-\chi^2)]^{1/2} \tag{6.3.9}$$

Here β is the cosine of the angle between vector r and the projection of ω on the plane x_1, x_2. The upper sign corresponds to the case $arccos\ \beta < \pi$ and lower to the case $arccos\ \beta > \pi$.

Expressions (6.3.8), (6.3.9) permit us to introduce different scales of turbulence as it was done in [37]. For small distances we have the following expansions:

$$Q_{ij} = a_1 r_i r_j - (a_2 + a_3 r^2)\delta_{ij} - \{b_1 +$$

$$[b_2 + (2b_2 + 3b_3)\chi^2]r^2\}(\omega_{im} r_j + \omega_{jm} r_i) r_m \tag{6.3.10}$$

$$+ r\chi[5b_4 + 7(b_1 + b_3\chi^2)r^2](\omega_{im}\omega_j + \omega_{jm}\omega_i) r_m$$

Equation (6.3.6) corresponds to a diffusion process in the *(n+1)th* space *at a finite stage of dissipation,* because inertia terms are negligible:

$$\frac{\partial Q}{\partial t} = 2\nu\Delta_n Q,$$

$$Q(t = 0) = Q^0(r, \chi), \tag{6.3.11}$$

$$\lim Q = 0\ (r \to \infty), \qquad n = 2, 4, 6$$

The solution of (6.3.11) can be expressed through the Green function of the diffusion equation as

$$Q(r,\chi,t) = \frac{1}{(8\pi\nu t)^{(n+1)/2}} \int \exp\left(-\frac{\left|r - r'\right|^2}{8\nu t}\right) Q^0(r',\chi)dr' \tag{6.3.12}$$

Let us introduce polar coordinates, for example, in the space of seven variables:

$$x_1 = r\cos\theta, \qquad x_2 = r\sin\theta\cos\varphi_1,$$

$$\tag{6.3.13}$$

$$x_3 = r\sin\theta\ \ \sin\varphi_1\cos\varphi_2$$

$$x_4 = r \sin\theta \ \sin\varphi_1 \sin\varphi_2 \cos\varphi_3,$$

$$x_5 = r \sin\theta \ \sin\varphi_1 \sin\varphi_2 \sin\varphi_3 \cos\varphi_4,$$

$$x_6 = r \sin\theta \ \sin\varphi_1 \sin\varphi_2 \sin\varphi_3 \sin\varphi_4 \cos\varphi_5,$$

$$x_7 = r \sin\theta \ \sin\varphi_1 \sin\varphi_2 \sin\varphi_3 \sin\varphi_4 \sin\varphi_5 \cos\varphi_6,$$

The rule of coordinate construction in the space of other dimension is now clear.

Let us assume that vector $\boldsymbol{r}$ is in the plane x_1, x_2 and $\cos\theta = \chi$, that is,

$$|\boldsymbol{r} - \boldsymbol{r}'|^2 = r^2 + r'^2 - 2r\,r'\cos\beta,$$

$$\cos\beta = \chi\chi' + [(1-\chi^2)(1-\chi'^2)]^{1/2}\cos\varphi_1'. \tag{6.3.14}$$

Expressing the volume element $dr_n = dx_1$ ...through dr, $d\theta$, $d\varphi_1$, ...and integrating over all $d\varphi_k$ $(k > 1)$, we transform (6.3.12) to the following form:

$$Q(r,\chi,t) = \frac{b_n''}{(8\pi\nu t)^{(n+1)/2}} \int_0^\infty\int_0^\pi\int_0^\pi \exp\left(-\frac{r^2 + r'^2 - 2rr'\cos\beta}{8\nu t}\right) \times$$

$$\times\, Q^0(r',\chi')r'^n \sin^{n-1}\theta' \sin^{n-2}\varphi_1' dr' d\theta' d\varphi_1' \tag{6.3.15}$$

Here b'' is 1, 4π or $(4/3)\pi^2$, correspondingly, at $n = 2$, 4 and 6.

This integral will be essentially simpler if one introduces the extension through the Gegenbauer polynomials.

The theorem for the Bessel functions gives:

$$e^{\gamma\cos\beta} = 2^\lambda\Gamma(\lambda)\sum_{m=0}^{\infty}(m+\lambda)C_m^\lambda(\cos\beta)\frac{I_{m+\lambda}(\gamma)}{\gamma^\lambda} \tag{6.3.16}$$

where $\Gamma(x), I_m(x), C_m^\lambda(x)$ are gamma-functions, the Bessel function of imaginary argument and the Gegenbauer polynomial, correspondingly.

Here we assume that λ is not equal to zero or negative integer number.

Let us use the well-known integral:

$$\int_0^n C_m^\lambda(\cos\beta)\sin^{2\lambda-1}\varphi_1' d\varphi_1' = \frac{2^{2\lambda-1}\Gamma^2(\lambda)m!}{\Gamma(2\lambda+m)} C_m^\lambda(\chi)C_m^\lambda(\chi') \tag{6.3.17}$$

Selecting value λ equal to $\lambda_n=(n-1)/2$ for the space of dimension $n+1$, we obtain:

$$Q(r,\chi,t) = \frac{b_n'}{(8\pi\nu t)^{(n+1)/2}} \exp(-\frac{r^2}{8\nu t}) \sum_{m=0}^{\infty} \frac{(m+\lambda_n)m!}{(m+n-2)!} C_m^{\lambda_n}(\chi) \times$$

$$\times \int\int \exp\left(-\frac{r'^2}{8\nu t}\right) C_m^{\lambda_n}(\chi') \frac{I_{m+\lambda_n}(\gamma)}{\gamma^{\lambda_n}} Q^0(r',\chi') r'^n (1-\chi'^2)^{(n/2)-1} dr' d\chi' \tag{6.3.18}$$

Here:

$$b_n'' = b_n' 2^{3\lambda_n-1}\Gamma^3(\lambda_n), \quad \gamma = \frac{rr'}{4\nu t}$$

Let us assume analogously to [41] that $Q^0(r,\chi)$ can be expanded through the Gegenbauer polynomials:

$$Q^0(r.\chi) = \sum q_m^0(r) C_m^\lambda(\chi) \tag{6.3.19}$$

Using the orthogonal relation for these polynomials:

$$\int_{-1}^{1} C_m^\lambda(\chi) C_{m'}^\lambda (1-\chi^2)^{\lambda-\frac{1}{2}} d\chi = \delta_{mm'} \frac{2^{1-2\lambda}\pi\Gamma(m+2\lambda)}{\Gamma^2(\lambda)(m+\lambda)m!} \tag{6.3.20}$$

We obtain finitely the solution of the problem (6.3.11) as

$$Q_1(r,t) = \frac{b_n}{(\nu t)^{(n+1)/2}} \exp\left(-\frac{r^2}{8\nu t}\right) \sum_{m=0}^{\infty} C_m^{\lambda_n}(\chi) \times$$

(6.3.21)

$$\times \int \exp\left(-\frac{r^2}{8\nu t}\right) \frac{I_{m+\lambda_n}(\gamma)}{\gamma^{\lambda_n}} q_m^0(r') r'^n dr'; \qquad \gamma = \frac{rr'}{4\nu t};$$

$$b_n = 2^{-(n+2)} \pi^{-\lambda_n} \Gamma(\lambda) b_n'', \qquad b_2'' = 1,$$

$$b_4'' = 4\pi, \qquad b_6'' = \frac{4}{3}\pi^2; \qquad \lambda_n = \frac{1}{2}(n-2)$$

Now we can find the expressions of Q_k according to (6.2.13) accounting for independence of Q_1 on χ and that Q_2 is an odd function of χ. Then:

$$Q_1(r,t) = \frac{(\nu t)^{-5/2}}{32} \int_0^{\infty} \exp\left(-\frac{r'^2 + r^2}{8\nu t}\right) \frac{I_{3/2}(\gamma)}{\gamma^{3/2}} Q_1^0(r') r'^4 dr';$$

$$Q_2(r,\chi,t) = \frac{1}{256(\nu t)^{7/2}} \exp\left(-\frac{r^2}{8\nu t}\right) \sum_{m=1}^{\infty} C_{2m-1}^{5/2}(\chi) \times \qquad (6.3.22)$$

$$\int_0^{\infty} \exp\left(-\frac{r'^2}{8\nu t}\right) \frac{I_{2m+3/2}(\gamma)}{\gamma^{5/2}} q_{2,2m-1}^0(r') r'^6 dr'.$$

Here $q^0_{2,2m-1}(r)$ – the coefficient of the expansion of $Q^0_2(r,\chi)$ through the Gegenbauer polynomials with χ of odd order.

Consider asymptotic expressions of these solutions at $t \to \infty$. Using the Bessel function definition as the set, we obtain:

$$Q_1(r, t) = \frac{1}{48(2\pi)^{1/2}(\nu t)^{5/2}} \exp(-\frac{r^2}{8\nu t}),$$

$$\Lambda_1 = \int_0^\infty r^4 Q_1^0(r)dr \qquad (6.2.23)$$

$$Q_2(r, \chi, t) = \frac{1}{3072(2\pi)^{1/2}(\nu t)^{9/2}} \exp(-\frac{r^2}{8\nu t}),$$

$$\Lambda_2 = \int_0^\infty r^7 q_{2,1}^0(r)dr$$

The first expression is a well-known result for isotropic turbulence ($\Lambda_1$is the Loitzanskiy [169] invariant). The second one corresponds to vortex anisotropy.

The tensor Q_{ij} has the same form at the decay stage as in the isotropic case:

$$Q_{ij}(r,t) \approx \frac{1}{24(2\pi)^{1/2}(\nu t)^{5/2}} \exp(-\frac{r^2}{8\nu t})\,\Lambda_1 \delta_{ij} \qquad (6.2.24)$$

(Here we hold the terms of the lowest order by t^{-1}.)

6.4. TURBULENCE WITH ANGULAR SYMMETRY

We shall look for dependency of these functions on spin angular velocity ω_l . This means dependency on a pseudo-vector, that is, we consider averaged parameters *without a mirror type of symmetry.*

Let us neglect the Batchelor condition (6.3.1) that excluded rotation symmetry. Then the antisymmetrical potentials are not zero [36, 37] and the following results are valid for potentials of the tensor Γ_{ij}:

$$\Gamma_3 = \Delta_4 Q_3, \qquad \Gamma_4 = \Delta_2 Q_4 + 2D_\chi Q_3 \qquad (6.4.1)$$

Therefore potentials Q_3, Q_4 will satisfy the following equations:

$$\frac{\partial Q_3}{\partial t} = 2\nu\Delta_4 Q_3 + S_3,$$

$$\frac{\partial Q_4}{\partial t} = 2\nu(\Delta_2 Q_4 + \frac{2}{r}\frac{\partial Q_3}{\partial \chi}) + S_4 \tag{6.4.2}$$

The solution of the first equation (6.4.2) has the form:

$$Q_3(r, \chi, t) = \frac{1}{32(\nu t)^{3/2}} \exp\left(-\frac{r^2}{8\nu t}\right) \sum_{m=0}^{\infty} C_{2m}^{3/2}(\chi) \times$$

$$\times \int_0^{\infty} \exp\left(-\frac{r^2}{8\nu t}\right) \frac{I_{2m+3/2}(\gamma)}{\gamma^{\lambda_n}} q_{3,2m}^0(r')r'^4 dr'; \qquad \gamma = \frac{rr'}{4\nu t}; \tag{6.4.3}$$

The solution of the second equation of (6.4.2) is the sum of general solutions of (6.3.11) at n=2 and the specific solution. That is,

$$Q_4(r, \chi, t) = \frac{1}{16(\nu t)^{3/2}} \exp\left(-\frac{r^2}{8\nu t}\right) \sum_{m=1}^{\infty} C_{2m-1}^{3/2}(\chi) \times \tag{6.4.4}$$

$$\times \int_0^{\infty} \exp\left(-\frac{r'^2}{8\nu t}\right) \frac{I_{2m-1/2}(\gamma)}{\gamma^{\lambda_n}} q_{4,2m-1}^0(r')r'^2 dr' +$$

$$+ 4\nu \int_0^t \frac{dt'}{[8\pi\nu(t-t')]^{3/2}} \int \left(\frac{1}{r^2}\frac{\partial Q_3}{\partial \chi}\right)_{r', \chi', t'} \exp\left[-\frac{\left|r - r'\right|^2}{8\nu(t-t')}\right] dr_2'$$

The asymptotic representations of (6.4.3) and (6.4.4) will have the following form:

$$Q_3(r, \chi, t) \approx \frac{\Lambda_3}{48(2\pi)^{1/2}(\nu t)^{5/2}} \exp(-\frac{r^2}{8\nu t}), \tag{6.4.5}$$

$$\Lambda_3 = \int_0^\infty r^4 q^0_{3,0}(r)dr$$

$$Q_4(r, \chi, t) \approx \frac{\Lambda_4 r\chi}{96(2\pi)^{1/2}(\nu t)^{5/2}} \exp(-\frac{r^2}{8\nu t}),$$

$$\Lambda_4 = \int_0^\infty r^3 q^0_{4,1}(r)dr$$

At the last stage of decay we have the approximation:

$$Q_{ij}(r, t) \approx \frac{1}{24(2\pi)^{1/2}(\nu t)^{5/2}} \exp(-\frac{r^2}{8\nu t}) \times$$

$$\times [\Lambda_1\delta_{ij} + \frac{1}{2}(\Lambda_3 + \frac{1}{2}\Lambda_4)\omega_{ij}] \tag{6.4.6}$$

Formula (6.2.24) for Q_{ij} is valid if $\Lambda_4 = -2\Lambda_3$.

Consider the case when the Kolmogorov structure function [138] can be represented by the pulsation field:

$$D_{ij} = \overline{(w_i - w_i')(w_j - w_j')} \tag{6.4.7}$$

We can see that for small distances between points M and M' the structure function corresponds exactly to correlations of the velocity gradient pulsation, that is, of distortion rates:

$$D_{ij} \approx \overline{\frac{\partial w_i}{\partial x_k}\frac{\partial w_j}{\partial x_l}} r_k r_l = D_{ijkl} r_k r_l$$

The distortion tensor can be represented as a sum of symmetric and skew tensors:

$$\frac{\partial w_i}{\partial x_k} = e'_{ij} + \omega_{ij}$$

Assume that the mean value of antisymmetric tensor $\omega_{ij} = \varepsilon_{ijk}\ \omega_k$ is the only parameter determining the turbulent structure of fluid flow. The local deformation rate e_{ij} will be included just into the dissipation.

The fourth order tensor D_{ijkl} does not depend on radius r_k and is symmetric to the coupled indexes commutation ($D_{kl\,ij} = D_{ij\,kl}$). The general view of this tensor is the following:

$$\begin{aligned} D_{ijkl} &= A\delta_{ij}\delta_{kl} + B_1\delta_{kj}\delta_{li} + B_2\delta_{ik}\delta_{lj} + \\ &C(\omega_{ik}\delta_{lj} + \omega_{jl}\delta_{ik}) + D(\omega_{il}\delta_{kj} + \omega_{jk}\delta_{il}) + \\ &E(\omega_{il}\omega_j\omega_k + \omega_{jk}\omega_l\omega_i) + K_1\omega_{ij}\omega_{lk} + \\ &F(\omega_{ik}\omega_j\omega_l + \omega_{jl}\omega_k\omega_i) + K_2\omega_{il}\omega_{jk} + K_3\omega_{jl}\omega_{ik}. \end{aligned} \tag{6.4.8}$$

Because $D_{ik\,jl} \neq D_{il\,jk}$ we have, generally speaking, $B_1 \neq B_2$; $K_1 \neq K_2 \neq K_3$.

Let us consider now the limitations on the tensor D_{ijkl} that follows from the continuity equation:

$$\frac{\partial w_1}{\partial r_1} = -\left(\frac{\partial w_2}{\partial r_2} + \frac{\partial w_3}{\partial r_3}\right). \tag{6.4.9}$$

We have:

$$\overline{\left(\frac{\partial w_1}{\partial r_1}\right)^2} = \overline{\left(\frac{\partial w_2}{\partial r_2} + \frac{\partial w_3}{\partial r_3}\right)\frac{\partial w_1}{\partial r_1}}, \dots \tag{6.4.10}$$

Summation of the three such equalities gives us the connection used earlier by Taylor [279]:

$$\overline{\left(\frac{\partial w_1}{\partial r_1}\right)^2} + \overline{\left(\frac{\partial w_2}{\partial r_2}\right)^2} + \overline{\left(\frac{\partial w_3}{\partial r_3}\right)^2} = . \qquad (6.4.11)$$

$$-\,2\left(\overline{\frac{\partial w_1}{\partial r_1}\frac{\partial w_2}{\partial r_2}} + \overline{\frac{\partial w_2}{\partial r_2}\frac{\partial w_3}{\partial r_3}} + \overline{\frac{\partial w_1}{\partial r_1}\frac{\partial w_3}{\partial r_3}}\right).$$

Introducing here the components of a D-tensor, we get $3(A+B_1+B_2)= -6B_1$ and, finally,

$$B_1 = B_2 = B.$$

Further, multiplication of the continuity equation and $\partial w_1 / \partial r_2$ after averaging will give:

$$D_{1112} + D_{2212} + D_{3312} = 0. \qquad D_{1121} + D_{2221} + D_{3321} = 0.$$

Summarizing these two equations, and accounting for $\omega_{21} = -\omega_{12}$, we have:

$$(K_1 - K_2)\,\omega_{31}\,\omega_{32} = 0, \quad K_2 = K_3 = -K_1, \quad K_1 = K.$$

On the contrary, subtracting one of the equations from the other, we get:

$$3C + 2D + E + 2F(\omega_1^2 + \omega_2^2) = 0$$

Analogously, we can get also two other equations:

$$3C + 2D + E + 2F(\omega_2^2 + \omega_3^2) = 0; \quad 3C + 2D + E + 2F(\omega_3^2 + \omega_1^2) = 0;$$

Hence, we have: $3(3C + 2D) + 3E + 4F = 0, \qquad F = 0$. Therefore, $E = -10B$, $D = -4B$, $C = 6B$. Finally, the D-tensor has the following form:

$$D_{ijkl} = A\delta_{ik}\delta_{jl} - \frac{A}{4}(\delta_{kj}\delta_{li} + \delta_{ik}\delta_{lj}) +$$

$$C(\omega_{ij}\delta_{kl} + \omega_{kl}\delta_{jk}) - \frac{2}{3}C(\omega_{il}\delta_{ij} + \omega_{jl}\delta_{ik}) -$$

$$\frac{5}{3}C(\omega_{ik}\omega_j\omega_l + \omega_{jl}\omega_k\omega_i) + K(\omega_{ij}\omega_{lk} - \omega_{il}\omega_{jk} - \omega_{jl}\omega_{ik}).$$

Therefore, the correlation tensor of local velocity gradients includes three independent scalars.

Let us express the first one through the energy dissipation rate:

$$A = \frac{15}{2\nu\rho}W; \qquad W = \nu\rho\overline{\left(\frac{\partial w_i}{\partial r_j} + \frac{\partial w_j}{\partial r_i}\right)^2}.$$

Constant K can be expressed through the vorticity correlation $\overline{\omega_i\omega_j}$. In fact,

$$\overline{\omega_i\omega_j} = \frac{1}{2}\varepsilon_{ilk}\varepsilon_{jmn}\overline{\frac{\partial w_k}{\partial x_l}\frac{\partial w_n}{\partial x_m}}$$

and

$$2K = \frac{1}{4}\frac{W}{\rho\nu} - < \omega^2 > .$$

In the isotropy case $W=4\rho\nu\overline{\omega^2}$ and $K = 0$.

Constant C appears in the interconnections between correlations of stretching and shear rates:

$$< \frac{\partial w_1}{\partial r_2}\frac{\partial w_2}{\partial r_1} > = \frac{A}{4}\left[\overline{\left(\frac{\partial w_1}{\partial r_2}\right)^2}\;\overline{\left(\frac{\partial w_2}{\partial r_1}\right)^2}\right]^{\frac{1}{2}}\left[A^2 - (\frac{5}{3}C\omega_1\omega_2\omega_3)^2\right]^{-\frac{1}{2}}.$$

If one of the components of pseudovector $\boldsymbol{\omega}$ is equal to zero, the latter equation is transformed into the Taylor expression for the isotropic case ($\omega_l = 0$):

$$\overline{\frac{\partial w_1}{\partial r_2}\frac{\partial w_2}{\partial r_1}} = \frac{1}{4}\left[\overline{\left(\frac{\partial w_1}{\partial r_2}\right)^2}\;\overline{\left(\frac{\partial w_2}{\partial r_1}\right)^2}\right]^{\frac{1}{2}},$$

$$\frac{1}{2}\overline{\left(\frac{\partial w_1}{\partial r_2}\right)^2} = \left(\frac{10C}{3\,A}\omega_1\omega_2\omega_3 - 1\right)\overline{\left(\frac{\partial w_1}{\partial r_1}\right)^2}, \quad \frac{1}{2}\overline{\left(\frac{\partial w_1}{\partial r_3}\right)^2} = \left(\frac{10\,C}{3\,A}\omega_1\omega_2\omega_3 - 1\right)\overline{\left(\frac{\partial w_1}{\partial r_1}\right)^2}$$

.

CHAPTER 7

VECTOR - DIRECTOR CONCEPT IN TURBULENCE

7.1. DYNAMICS EQUATIONS

In a close vicinity of channel walls, the mesostructure of turbulent flow is characterized by turning of vortex orientation. Probably a description of eddy transformation needs more sophisticated objects than a rotating "mole". Correspondingly, we have to nominate a proper type of anisotropy and find the parameter that could be dynamically changed in a flow. There have been different approaches taken to this problem. For example, the given anisotropy [175] was taken into account directly in the constitutive equations for the Reynolds stresses and strain rates. The Cosserat continuum is used in this book for turbulent currents with spin velocity of eddies as thermodynamic parameters [7, 118, 119, 222].

The vector-director approach, developed in the liquid crystal theory by E. Aero, J. Ericksen and F. Leslie [2, 77-79, 160] will be considered here[22], that is, we will use [14-16, 177-179] for description of the anisotropy evolution in space and time. Let us consider a turbulent fluid as the special continuum (in accordance with the Reynolds philosophy), in which local parameters are the result of space averaging. As it was shown above, a turbulent fluid possesses internal angular momentum which balance has to be added to the conventional balances of mass and impulse. In Cartesian coordinates X_i , for incompressible fluids they are obtained as (Chapters 2 and 3):

$$\frac{\partial U_\alpha}{\partial X_\alpha} = 0, \qquad (7.1.1)$$

$$\rho\frac{dU_i}{dt} = \frac{\partial \sigma_{i\alpha}}{\partial X_\alpha} + \rho F_i, \qquad (7.1.2)$$

[22] Vladimir A. Babkin modified this approach for the turbulence case. His results are given in this chapter.

$$\frac{dM_i}{dt} = \frac{\partial \mu_{i\alpha}}{\partial X_\alpha} + \varepsilon_{i\alpha\beta}\sigma_{\beta\alpha} + \rho\, C_i, \qquad (7.1.3)$$

$$\frac{d}{dt} = \frac{\partial}{\partial t} + U_\alpha \frac{\partial}{\partial X_\alpha}, \qquad (7.1.4)$$

Again ρ is the density, U_i is the velocity, σ_{ij} are the stresses, μ_{ij} are the couple stresses, F_i, C_i, are the external bulk force, external bulk moment and M_i is the angular momentum per unit volume, correspondingly.

For definition of the angular momentum M_i, let us consider the smallest elements of vortex structure as a number of k spherical eddies («turbulent mols» [240]), rotating around their own axis with angular velocity $\omega^{(k)}{}_i$. An eddy's own rotation also can rotate; the angular speed of an axis itself we shall designate $\Omega^{(k)}{}_i/2$. The angular moment of rotation of one eddy m_i is defined as

$$m_i = \rho\, j_{i\alpha}(\omega_{0\alpha} + \Omega_{0\alpha}), \qquad j_{ik} = j\delta_{ik},$$

Here j_{ik} is the moment of inertia of a separate eddy, δ_{ik} is a unit tensor.

Assume that a small vicinity of a macropoint (X_1, X_2, X_3) is the representative volume of a fluid ΔV with mass Δm.

The average angular moment of spin rotation of an average eddy is defined as

$$M_i^\omega = \frac{1}{\Delta m}\sum_{(k)} j^{(k)}\omega_{0i}^{(k)}, \qquad (7.1.5)$$

where summation will be carried out over all eddies in volume ΔV, marked index k.

By transition to a limit at $(\Delta V/ L^3) \rightarrow 0$, the variable M_i^ω will be determined (here L is the external linear scale as above).

The average moments of inertia are determined similarly by transition to the same limit:

$$J_{ik} = J\delta_{ik},$$

$$J = \frac{1}{\Delta m}\sum_{(k)} j^{(k)} \qquad (7.1.6)$$

From (7.1.5) and (7.1.6) we get

$$M_i^{\omega} = \rho J \omega_i ,$$
$$\omega_i = \frac{1}{J\,\Delta m} \sum_{(k)} j^{(k)} \omega_{0i}^{(k)} \tag{7.1.7}$$

Here, after the transition to a limit at $(\Delta V / L^3) \to 0$, the spin ω_i of own rotations of the average eddy appears.

The similar, referred to volume unit, angular moment of the eddies, caused by rotation of own axes with angular velocity $(1/2)\,\Omega_{0i}^{(k)}$, is represented as

$$M_i^{\Omega} = \rho J \Omega_i ,$$
$$\Omega_i = \frac{1}{J\Delta m} \sum_{(k)} j^{(k)} \Omega_{0i}^{(k)} \tag{7.1.8}$$

and, after transition to a limit at $(\Delta V / L^3) \to 0$, field variables Ω_i and M_i^{Ω} appear.

Let us introduce in the elementary volume ΔV (with the point X_i as a mass center) at each point of flow the unit vector ℓ_i, in parallel to which eddies' own axes of rotations are situated on average.

Assume that the eddy spin ω_i is equal to zero at each point of a flow.

As, by definition, $(1/2)\,\omega_i$ is an average spin of eddies in the elementary volume, containing densely located ascending and descending branches of a system of Λ-eddies [39, 105, 191, 233, 234], the rotations in which are opposite, the given assumption is justified.

It means that in each element of the turbulent medium the spin axes of eddies are on average parallel to the vector ℓ_i but the module of their average angular velocity is equal to zero. At $\omega_i = 0$ we have

$$M_i = \rho J \Omega_i \tag{7.1.9}$$

Then equation (7.1.3) has a form

$$\rho \frac{dJ\Omega_i}{dt} = \frac{\partial \mu_{i\alpha}}{\partial x_\alpha} + \varepsilon_{i\alpha\beta}\sigma_{\beta\alpha} + \rho C_i \tag{7.1.10}$$

If the pseudovector Ω_i and the vector ℓ_i are orthogonal to each other, then

$$\frac{d\ell_i}{dt} = \varepsilon_{i\alpha\beta}\Omega_\alpha \ell_\beta, \qquad \Omega_i = \varepsilon_{i\alpha\beta}\ell_\alpha \frac{d\ell_\beta}{dt}. \tag{7.1.11}$$

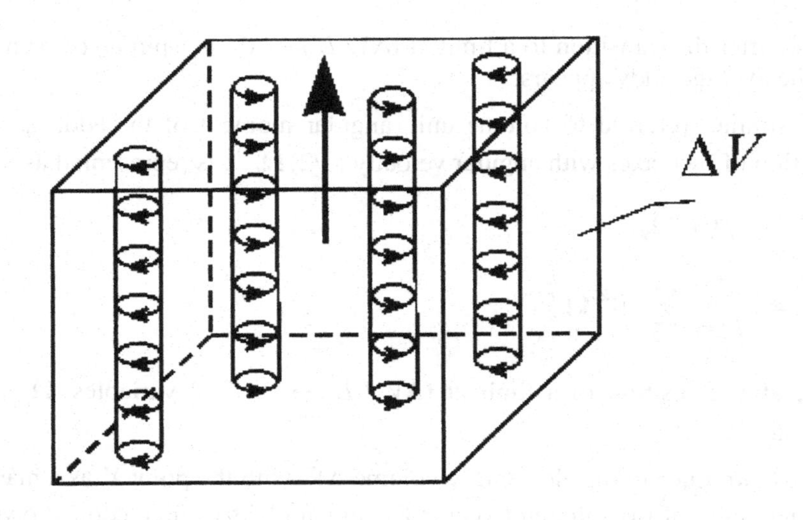

Figure 7.1.1. Although vortexes in ΔV can have zero total rotation, they correspond to some vector-director

Substituting the second formula of (7.1.11) into equation (7.1.10), we get

$$\varepsilon_{i\alpha\beta}\rho\ell_\alpha \frac{d}{dt}(J\frac{d\ell_\beta}{dt}) = \frac{\partial \mu_{i\alpha}}{\partial x_\alpha} + \varepsilon_{i\alpha\beta}\sigma_{\beta\alpha} + \rho C_i. \tag{7.1.12}$$

From equality (7.1.12) follows

$$\rho \frac{d}{dt}(J\frac{d\ell_i}{dt}) = \varepsilon_{i\alpha\beta}\left(\frac{\partial \mu_{\alpha k}}{\partial x_k} + \varepsilon_{\alpha\mu\nu}\sigma_{\nu\mu} + \rho G_\alpha\right)\ell_\beta + \Lambda\ell_i,$$

Here Λ is any scalar. This equation is possible to present as

$$\rho \frac{d}{dt}(J\frac{d\ell_i}{dt}) = \frac{\partial \beta_{ij}}{\partial x_j} + g_i + \rho C_i, \tag{7.1.13}$$

$$\beta_{ij} = \varepsilon_{i\alpha\beta}\mu_{\alpha j}\ell_{\beta}, \tag{7.1.14}$$

$$g_i = \lambda\ell_i + \varepsilon_{i\alpha\beta}(\varepsilon_{\alpha kv}\sigma_{vk}\ell_{\beta} - \mu_{\alpha k}\frac{\partial \ell_{\beta}}{\partial x_k}) \tag{7.1.15}$$

$$G_i = \varepsilon_{i\alpha\beta}C_{\alpha}C_{\beta}. \tag{7.1.16}$$

If $J = const$, equation (7.1.13) has a form

$$\rho J\frac{d^2\ell_i}{dt^2} = \frac{\partial \beta_{ij}}{\partial x_j} + g_i + \rho G_i. \tag{7.1.17}$$

Considering the conditions we accepted as valid for equation (7.1.17) was received as valid ones, we use this equation for turbulent flow and accept velocity U_i and vector ℓ_i as adequate kinematical parameters.

If $J \neq const$, a system of dynamics equations can be constructed by adding by the evolution equation for parameter J (as it was done above, see also [4,5]). Equation (7.1.17) is a typical equation of an *oriented fluid* [2, 78, 79, 160], in which the vector ℓ_i is named as the director [2].

Variables β_{ij}, g_i, G_i are known as generalized stresses, the generalized internal and generalized external mass forces [2]. J. Ericksen [77-79] and F. Leslie [160] postulated only equation (7.1.17) and its connection with the angular momentum balance was specified in [2]. Accepting equation (7.1.1), (7.1.2) and (7.1.17) as dynamics equations of the medium, we consider the latter as an oriented fluid, the detailed properties of which will be specified by the constitutive laws.

In 1989 the first paper by Marshall and Naghdi [177-179] were published where the model of oriented fluid was also suggested for the turbulence theory. We compare the Babkin approach [14] of 1985 with the latter. Note the following:

i). Kinematical parameters U_i and ℓ_i in both models are identical.

ii). Connection of the director with the main axis of the maximum main strain rate [39, 233] seems too restrictive. This means that in the case of velocity profiles as considered below the director has a constant inclination to the flow equal to 45^0.

Experimentally and analytically it was established [39, 105, 191, 233] that asymptotically far from the wall the inclination of the system of Λ-eddies to the flow direction in a plane case is really close to *45⁰*. However, near the wall the Λ-eddies are extended along flow lines and afterwards the directions of the Λ-eddies varies continuously.

In addition, according to [249], the maximal value of the inclination of the Λ-eddies decreases with growth of the Reynolds number. In [249] the values are observed in the interval: $35^0 - 40^0$.

However, if the mentioned condition is rejected, it is evident in section 7.4 below that solutions with constant length of the director are possible and they are in a good agreement with the experimental results. It is worth noting that actually density of Λ-eddies is decreasing with a distance from the solid wall [233, 234] and the flows with constant director length belong to some average situation, only approximately reflecting a true picture of the phenomenon.

7.2. ENERGY AND CONSTITUTIVE LAWS

Multiply equation (7.1.10) by the vector Ω_i in a scalar manner:

$$\rho\Omega_i \frac{dJ\Omega_i}{dt} = \Omega_i \frac{\partial \mu_{i\alpha}}{\partial X_\alpha} + \varepsilon_{i\alpha\beta}\Omega_i\sigma_{\beta\alpha} + \rho C_i\Omega_i . \tag{7.2.1}$$

The replacement of components Ω_i by expressions (7.1.11) with use of formulas (7.1.14) and (7.1.15) will transform equality (7.2.1) to a form

$$\rho \frac{d}{dt}\left(\frac{J\dot{\ell}_i\dot{\ell}_i}{2}\right) = \frac{\partial \beta_{i\alpha}\dot{\ell}_i}{\partial X_\alpha} - \beta_{i\alpha}\frac{\partial \dot{\ell}_i}{\partial X_\alpha} + g_i\dot{\ell}_i + \rho G_i\dot{\ell}_i, \tag{7.2.2}$$

$$\dot{\ell}_i = \frac{d\ell_i}{dt}. \tag{7.2.3}$$

Multiplication of equation (7.1.2) by U_i and subsequent summarizing with equation (7.2.2) will give the equation of balance of kinetic energy.

$$\rho\frac{d}{dt}\left(\frac{U_iU_i}{2} + \frac{J\dot{\ell}_i\dot{\ell}_i}{2}\right) = \frac{\partial(\sigma_{i\alpha}U_i)}{\partial X_\alpha} + \frac{\partial(\beta_{i\alpha}\dot{\ell}_i)}{\partial X_\alpha} - \sigma_{i\alpha}\frac{\partial U_i}{\partial X_\alpha} - \tag{7.2.4}$$

$$\beta_{i\alpha} \frac{\partial \dot{\ell}_i}{\partial X_\alpha} + \rho F_i U_i + g_i \dot{\ell}_i + \rho G_i \dot{\ell}_i$$

For a fluid volume *V*, limited by the surface *S*, equality (7.2.4) can be written in the integrated form

$$\frac{d}{dt} \int_V \rho \left(\frac{U_i U_i}{2} + \frac{J \dot{\ell}_i \dot{\ell}_i}{2} \right) dV =$$

$$\int_V (\rho F_i U_i + g_i \dot{\ell}_i + \rho_i G_i \dot{\ell}_i - \sigma_{ij} \frac{\partial U_i}{\partial X_j} - \beta_{ij} \frac{\partial \dot{\ell}_i}{\partial X_j}) dV + \qquad (7.2.5)$$

$$\int_S (\sigma_{ij} U_i + \beta_{ij} \dot{\ell}_i) dS_j$$

Here the cumulative rotation energy is accounted for by terms including rate of director changes $d\ell / dt$.

In the elementary volume *V* with surface *S* the external forces are mass force F_i, generalized mass force G_i and surface forces-stresses σ_{ij} and β_{ij}. Therefore, the first law of thermodynamics will be written as

$$\frac{d}{dt} \int_V \rho \left(\frac{U_i U_i}{2} + \frac{J \dot{\ell}_i \dot{\ell}_i}{2} + \varepsilon \right) dV = \int_V \rho (F_i U_i + G_i \dot{\ell}_i) dV$$

$$+ \int_S (\sigma_{ij} U_i + \beta_{ij} \dot{\ell}_i) dS_j + \int_V Q dV - \int_S q_j dS_j \qquad (7.2.6)$$

Here ε is the density of internal energy referred to mass unit, Q - heat source, q_i - heat flow.

From equations (7.2.5) and (7.2.6) the equation of heat balance follows:

$$\frac{d}{dt}\int_V \rho\,\varepsilon\, dV = \int_V \left(\sigma_{ij} \frac{\partial U_i}{\partial x_j} + \beta_{ij} \frac{\partial \dot{\ell}_i}{\partial x_j} - g_i \dot{\ell}_i \right) dV \tag{7.2.7}$$

$$+ \int_V Q dV - \int_S q_i dS_i$$

The transition to the local form gives the heat balance in the form of [160]

$$\rho \frac{d\varepsilon}{dt} = \sigma_{ij} e_{ij} + \beta_{ij} N_{ij} - g_i N_i + Q - q_{i,i} \tag{7.2.8}$$

where e_{ij}, N_i, N_{ij} are determined by the formulas

$$N_i = \dot{\ell}_i - \Phi_{i\alpha} \ell_\alpha, \qquad N_{ij} = \frac{\partial \dot{\ell}_i}{\partial X_j} - \Phi_{i\alpha} \frac{\partial \ell_\alpha}{\partial X_j} \tag{7.2.9}$$

$$2e_{ij} = U_{i,j} + U_{j,i}, \qquad 2\Phi_{ij} = U_{i,j} - U_{j,i}$$

Under the second law of thermodynamics for the oriented fluid model at $\ell = 1$ the following inequality [160] is valid:

$$\rho \frac{d\epsilon}{dt} - \frac{\partial}{\partial X_i}\left(\frac{q_i}{T}\right) - \frac{Q}{T} \geq 0, \tag{7.2.10}$$

where ϵ is the entropy of turbulent mass unit, T is the absolute temperature. If one uses the formula for free energy of turbulent and molecular chaos Ξ as a whole:

$$\mathfrak{I} = \varepsilon - T\,\epsilon,$$

then from equations (7.2.9) and (7.2.10) follows

$$\sigma_{ij}e_{ij} + \beta_{ij}N_{ij} - g_i N_i - \rho\left(\in \frac{dT}{dt} + \frac{d\Im}{dt}\right) - \frac{q_i}{T}\frac{\partial T}{\partial X_i} \geq 0 \quad (7.2.11)$$

Equality (7.2.11) is one of the basic expressions for selection of the constitutive equations. Formulation of the constitutive equations for σ_{ij}, β_{ij} and g_i, will be done analogously to the theory of nematic liquid crystals [2, 42, 78, 160]. Correspondingly,

$$\sigma_{ij} = -p\delta_{ij} + a_{ij} + \tau_{ij} \quad (7.2.12)$$

$$p = \rho^2 \frac{\partial \Im}{\partial \rho} \quad (7.2.13)$$

$$a_{ij} = -\rho \frac{\partial \Im}{\partial \ell_{\alpha,j}} \ell_{\alpha,i}, \quad (7.2.14)$$

$$\ell_{i,j} = \frac{\partial \ell_i}{\partial X_j}$$

$$\tau_{ij} = \mu_1 \ell_\alpha \ell_\beta e_{\alpha\beta} \ell_i \ell_j + \mu_2 N_i \ell_j + \mu_3 N_j \ell_i +$$

$$(7.2.15)$$

$$\mu_4 e_{ij} + +\mu_5 \ell_i \ell_\alpha e_{\alpha j} + \mu_6 \ell_j \ell_\alpha e_{\alpha i}$$

$$\beta_{ij} = \alpha_j \ell_i + \rho \frac{\partial \Im}{\partial \ell_{i,j}} \quad (7.2.16)$$

$$g_i = \ell_i + \Lambda_1 N_i + \Lambda_2 \ell_\alpha e_{i\alpha} - \rho \frac{\partial \Im}{\partial \ell_i} - \frac{\partial \alpha_\beta \ell_i}{\partial X_\beta} \quad (7.2.17)$$

$$\in = -\frac{\partial \Im}{\partial T} \quad (7.2.18)$$

Here p is pressure, a_{ij} are stresses, caused by vortex orientation in the flow, τ_{ij} are viscous stresses, also depending on mesostructure orientation Λ_1, Λ_2, μ_1, ..., μ_6

are constitutive parameters, γ and α_i are scalar and vector functions, ensuring the model's completeness.

Following [2, 16, 78], free energy $\Im$ shall be set as

$$2\rho\Im = 2\rho\Im_0 + k_{22}\ell_{\alpha,\beta}\ell_{\alpha,\beta} + k_{24}\ell_{\alpha,\beta}\ell_{\beta,\alpha} + (k_{11} - k_{22} - k_{24})\ell_{\alpha,\alpha}\ell_{\beta,\beta} + (k_{33} - k_{22})\ell_{\alpha}\ell_{\beta}\ell_{\gamma,\alpha}\ell_{\gamma,\beta} \tag{7.2.19}$$

where $\Im_0$ is the free energy of non-turbulized (isotropic) fluid, k_{11}, k_{22}, k_{33}, k_{24} are constants of the model. Factors Λ_1, Λ_2, μ_1, , μ_6, k_{11}, k_{22}, k_{33}, k_{24} are considered as dependent on flow structure, for example, on eddy density in a macropoint and on the mean module of their spin angular velocity. As this dependence demands further study, we shall consider them, for simplicity, as constants at the fixed global flow parameters (for example, at fixed Reynolds number). In case of an incompressible fluid, pressure p is determined by the dynamics equations.

The analysis shows that the factors of the oriented Ericksen-Leslie fluid model are interconnected in the following way [42]:

$$\begin{gathered} k_{11} \geq |k_{11} - k_{22} - k_{24}|, \quad k_{22} \geq |k_{24}|, \quad k_{33} \geq 0, \\ \Lambda_1 = \mu_2 - \mu_3, \quad \Lambda_2 = \mu_5 - \mu_6 = -(\mu_2 + \mu_3), \\ -4\Lambda_1(2\mu_4 + \mu_5 + \mu_6) \geq (\mu_2 + \mu_3 - \Lambda_2)^2, \quad \mu_4 \geq 0, \\ 2\mu_1 + 3\mu_4 + 2\mu_5 + 2\mu_6 \geq 0, \quad 2\mu_4 + \mu_5 + \mu_6 \geq 0. \end{gathered} \tag{7.2.20}$$

In the considered model of turbulent flow, condition (7.2.20) is also assumed to be valid.

The connection between the director field and angular momentum concept is explained. Particularly, if ϖ corresponds to rotation about the central axes parallel to the direction d_j, then

$$\Omega_j = \varepsilon_{jkl} d_k \dot{d}_l + d_j \varpi .$$

7.3. WIND VELOCITY PROFILE AT A GROUND SURFACE

As it was mentioned above, in a turbulent flow of a fluid along an infinite solid plane, the near-wall structure of longitudinal mass velocity looks like

$$\frac{U}{U_*} = \frac{1}{\kappa} \ln y + C, \tag{7.3.1}$$

where U is longitudinal velocity, y is the coordinate measured from a wall and perpendicular to it, $U_* = (\tau_w / \rho)^{½}$ is the dynamical velocity, τ_w is the modulus of shear stress at the plane, κ is the Karman constant, C is a constant.

Let us find conditions, under which the given model will give a profile of the type as in (7.3.1). So, assume that an incompressible fluid flows along an infinite solid plane.

Let axis x be in the direction of the flow and axis y be perpendicular to the plane. Projections of vectors U_i and ℓ_i will be found [16] as

$$U_x = U(y), \quad U_y = U_z = 0, \tag{7.3.2}$$

$$\ell_x = \cos \theta(y), \quad \ell_y = \sin \theta(y), \quad \ell_z = 0, \tag{7.3.3}$$

where $\theta(y)$ – the angle between the director and axis x.

The distribution (7.3.2) accounts for the mass balance. If one neglects the external mass forces F_i, G_i and pressure gradient, equation (7.1.2) and (7.1.17) can be written - in view of formula (7.2.12) - as:

$$\frac{d(a_{xy} + \tau_{xy})}{dy} = 0, \tag{7.3.4}$$

$$\frac{d(-p + a_{yy} + \tau_{yy})}{dy} = 0$$

$$\frac{d\beta_{xy}}{dy} + g_x = 0 \tag{7.3.5}$$

$$\frac{d\beta_{yy}}{dy} + g_y = 0$$

The substitution of component U_i and ℓ_i - in accordance with (7.3.2) and (7.3.3) - into equations (7.2.13) - (7.2.17), (7.2.19) gives an obvious expression for variables included in equation (7.3.4) and (7.3.15):

$$a_{xy} = 0, \quad , \qquad a_{yy} = -(k_{11}\cos^2\theta + k_{33}\sin^2\theta)\,\theta'^2 \tag{7.3.6}$$

$$2\tau_{xy} = \begin{bmatrix} 2\mu_1 \sin^2\theta\cos^2\theta + (\mu_5 - \mu_2)\sin^2\theta + \\ (\mu_3 + \mu_6)\cos^2\theta + \mu_4 \end{bmatrix} U' \tag{7.3.7}$$

$$2\tau_{yy} = \sin\theta\cos\theta(2\mu_1\sin^2\theta + \mu_2 + \mu_3 + \mu_5 + \mu_6)\,U'$$

$$\beta_{xy} = \alpha_y\cos\theta - \left[k_{22}\sin\theta + (k_{33} - k_{22})\sin^3\theta\right]\theta', \qquad \beta_{yy} = \alpha_y\sin\theta + \left[k_{11}\cos\theta + (k_{33} - k_{22})\cos\theta\sin^2\theta\right]\theta' \tag{7.3.8}$$

$$g_x = \gamma\cos\theta + \alpha_y\theta'\sin\theta + \tfrac{1}{2}(\Lambda_2 - \Lambda_1)\,U'\sin\theta,$$

$$g_y = \gamma\sin\theta - \alpha_y\theta'\cos\theta - (k_{33} - k_{22})\,\theta'^2\sin\theta + \tfrac{1}{2}(\Lambda_1 + \Lambda_2)\,U'\cos\theta \tag{7.3.9}$$

The prime designates derivative with respect to y; γ and α_2 are considered as constants.

After substitution of expressions (7.3.8) and (7.3.9) into equations (7.3.5) the constants γ and α_2, as appears, can be excluded. Then the following equation is obtained:

$$\frac{d}{d\theta}\left[(k_{11}\cos^2\theta + k_{33}\sin^2\theta)\theta'^2\right] + (\Lambda_1 + \Lambda_2\cos 2\theta)U' = 0. \qquad (7.3.10)$$

Assuming, as usual, that near solid wall $\tau_{xy} = \tau_w$, the integration of the first equation (7.3.4) will lead to

$$g(\theta)\, U' = \tau_w, \qquad (7.3.11)$$

we derive

$$2g(\theta) = 2\mu_1 \sin^2\theta \cos^2\theta + \qquad (7.3.12)$$

$$(\mu_5 - \mu_2)\sin^2\theta + (\mu_3 + \mu_6)\cos^2\theta + \mu_4$$

The exception by derivative U' from equation (7.3.10) and (7.3.11) results in the differential equation for definition of the angle θ:

$$\frac{d}{d\theta}\left[(k_{11}\cos^2\theta + k_{33}\sin^2\theta)\left(\frac{d\theta}{dy}\right)^2\right] = -\frac{\Lambda_1 + \Lambda_2\cos 2\theta}{g(\theta)}\tau_w. \qquad (7.3.13)$$

If velocity $U(y)$ is determined by formula (7.3.1), then, after finding U' and substituting it in equation (7.3.11), we get the algebraic equation for the angle θ:

$$g(\theta) = \rho\kappa U_* y \qquad (7.3.14)$$

Equality (7.3.14), in essence, is the integral of equation (7.3.13). The differentiation of (7.3.14) by y gives:

$$\frac{d\theta}{dy} = \frac{2\rho\kappa U_*}{(2\mu_1 \cos 2\theta - \mu_2 - \mu_3 + \mu_5 - \mu_6)\sin 2\theta}. \tag{7.3.15}$$

If one substitutes derivative (7.3.15) into equation (7.3.13), it is easy to see that the left side of the obtained expression is an odd, and the right side an even function of θ.

So, for equation (7.3.13) to be satisfied identically by substitution of derivative (7.3.15), it is necessary and sufficient that both - the left and the right sides - have to be zero. From this follows

$$k_{11} = k_{33} = 0, \qquad \Lambda_1 = \Lambda_2 = 0. \tag{7.3.16}$$

Taking into account for (7.3.16), expression (7.2.20) yields

$$\mu_2 = \mu_3 = 0, \qquad \mu_5 = \mu_6, \qquad k_{24} = -k_{22}. \tag{7.3.17}$$

Equalities (7.3.16) and (7.3.17) establish the restrictions that have to be satisfied by the constitutive coefficients for validity of the logarithmic velocity profile. Because we accept the Ericksen-Leslie model, its constants will satisfy conditions (7.3.16) and (7.3.17) besides conditions (7.2.20).

The study of the *plane turbulent flow of an incompressible flow the vicinity of a wall*, performed by Z. Zhang and K. Eisele [306], has established and experimentally confirmed the following:

(i). Turbulent (Reynolds) stresses at a point of a stationary flow R_{11}, R_{22}, R_{12} (Cartesian coordinates: X_1, X_2) are interconnected by the equality

$$R_{12} = \tfrac{1}{2}(R_{11} - R_{22})\,\mathrm{tg}\, 2\psi \tag{7.3.18}$$

Here ψ is the angle of an inclination of one of the stress tensor main axes R_{ij} to axis X_1.

(ii). If (X_1', X_2') is a Cartesian coordinate system rotated relative to the system (X_1, X_2) by the angle φ, turbulent stresses R_{11}, R_{12} in the new coordinates are related to the stresses in the old coordinates by the equality

$$R_{1'1'} = \frac{1}{2}(R_{11} + R_{22}) + \frac{\cos 2(\varphi - \psi)}{2\cos 2\psi}(R_{11} - R_{22}),$$

$$R_{1'2'} = -\frac{\sin 2(\varphi - \psi)}{2\cos 2\psi}(R_{11} - R_{22}). \tag{7.3.19}$$

Let us find out, at what level of the turbulent stresses the considered model is in accordance with the mentioned facts. For the plane flow case with the unit director we have

$$U_1 = U_1(X_1, X_2), \quad U_2 = U_2(X_1, X_2),$$
$$\ell_1 = \cos\theta(X_1, X_2), \quad \ell_2 = \sin\theta(X_1, X_2) \tag{7.3.20}$$

As equalities (7.3.18) and (7.3.19) relate the stresses R_{ij}, caused by turbulence, the formulas for stresses of the given model, also designated as R_{ij}, we shall take into account only those terms, in which the director and its derivatives clearly enter. Then, accounting for (7.3.16) and (7.3.17), equation (7.2.14), (7.2.15) and (7.2.19) will give:

$$R_{ij} = a_{ij} + \tau_{ij} - \mu_4 e_{ij} = k_{22}\frac{\partial \ell_\alpha}{\partial X_i}\left(\frac{\partial \ell_j}{\partial X_\alpha} - \frac{\partial \ell_\alpha}{\partial X_j} + \ell_j \ell_\beta \frac{\partial \ell_\alpha}{\partial X_\beta}\right)$$
$$+ \mu_1 \ell_\alpha \ell_\beta e_{\alpha\beta} \ell_i \ell_j + \mu_5 \ell_\alpha (\ell_i e_{\alpha j} + \ell_j e_{\alpha i}). \tag{7.3.21}$$

The substitution of functions (7.3.20) in formula (7.3.21) gives the evident expressions for R_{ij} (fluid is incompressible):

$$R_{11} = \mu_1 R(\theta)\cos^2\theta + 2\mu_5 \cos\theta\,(e_{11}\cos\theta + e_{12}\sin\theta)$$

$$R_{22} = \mu_1 R(\theta)\sin^2\theta + 2\mu_5 \sin\theta(e_{22}\sin\theta + e_{12}\cos\theta) \tag{7.3.22}$$

$$2R_{12} = 2R_{21} = \mu_1 R(\theta)\sin 2\theta + 2\mu_5 e_{12}$$

$$R(\theta) = e_{11}\cos^2\theta + e_{12}\sin 2\theta + e_{22}\sin^2\theta$$

After simple transformations formulas (7.3.22) yield the equality

$$2R_{12} = (R_{11} - R_{22})\mathrm{tg}\, 2\theta + 2\mu_5(e_{12} - e_{11}\mathrm{tg}\, 2\theta) \tag{7.3.23}$$

that coincides with (7.3.18) if one identifies the angles ψ and θ and puts $\mu_5 = 0$. Under the same conditions, stresses R_{11}, R_{22}, R_{12}, determined by formula (7.3.22), satisfy also equality (7.3.19), and this is easily established by direct checking.

In conclusion, taking into account all of the restrictions that we have accepted here on the model's coefficients, *the constitutive equations of turbulent flow in a wall layer* can be written [16] as

$$a_{ij} = k_{22}\frac{\partial \ell_\alpha}{\partial X_i}\left(\frac{\partial \ell_j}{\partial X_\alpha} - \frac{\partial \ell_\alpha}{\partial X_j} + \ell_j \ell_\beta \frac{\partial \ell_\alpha}{\partial X_\beta}\right) \tag{7.3.24}$$

$$\tau_{ij} = \mu_1 \ell_\alpha \ell_\beta e_{\alpha\beta} \ell_i \ell_j + \mu_4 e_{ij} \tag{7.3.25}$$

$$\beta_{ij} = \alpha_j \ell_i + k_{22}\left(\frac{\partial \ell_i}{\partial X_j} - \frac{\partial \ell_j}{\partial X_i} - \ell_j \ell_\alpha \frac{\partial \ell_i}{\partial X_\alpha}\right) \tag{7.3.26}$$

$$g_i = \gamma\ell_i - \frac{\partial(\alpha_\beta \ell_i)}{\partial X_\beta} + k_{22}\ell_\alpha \frac{\partial \ell_\beta}{\partial X_\alpha}\frac{\partial \ell_\beta}{\partial X_i} \tag{7.3.27}$$

One can see that in this case stress tensors are symmetric - due to (7.3.18) - and the eddy dynamics is accounted for by their nonlinear dependencies on the director ℓ_i.

To show the model's applications, we shall consider a number of classical problems. In all cases fluid is incompressible, flow is stationary and there are no mass forces. The system of Cartesian coordinates *x, y, z* will be used.

Structure of a wind at ground surface[23]

Neglecting mass forces F_i and G_i, speed U_i, vector - director ℓ_i and pressure we shall search for

$$U_x = U(y), \;\; U_y = U_z = 0, \;\; p = p(y), \tag{7.3.28}$$

$$\ell_x = \cos\theta(y), \; \ell_y = 0, \; \ell_z = \sin\theta(y)$$

Then the incompressibility equation is satisfied identically and equations (7.3.4) and (7.3.5) give the following values of the unknown variables

$$a_{xy} = a_{yy} = 0 \tag{7.3.29}$$

$$\tau_{xy} = (\mu_1 \sin^2\theta \cos^2\theta + \tfrac{1}{2}\mu_4)U',$$

$$\tau_{yy} = (\mu_1 \sin^3\theta \cos\theta)\, U', \tag{7.3.30}$$

$$\beta_{xy} = \alpha_y \cos\theta - k_{22}\,\theta' \sin\theta \cos^2\theta,$$

$$\beta_{yy} = \alpha_y \sin\theta - k_{22}\,\theta' \sin^2\theta \cos\theta, \tag{7.3.31}$$

$$g_x = \gamma \cos\theta + \alpha_y\,\theta' \sin\theta,$$

$$g_y = \gamma \sin\theta - \alpha_y\,\theta' \cos\theta + k_{22}\,\theta'^2 \sin\theta. \tag{7.3.32}$$

where α_y is the component of a vector α_i on an axis y and prime means its derivative by z. The coefficients α_y, γ, μ_1, μ_4, k_{22} in the formulas (7.3.30) - (7.3.32) are considered as constants.

Integrating the first equation of (7.3.4) together with the first formulas (7.3.29) and (7.3.30), we obtain

[23] Babkin V. A. On wind profile in a ground atmosphere layer (in press).

$$(\mu_1 \sin^2\theta \cos^2\theta + \frac{\mu_4}{2}) U' = \tau_w \qquad (7.3.33)$$

where τ_w is the module of a tangential stress at the ground surface.
The integration of the second equation (7.3.4) yields

$$p(y) = p(0) + \tau_{yy}(y) - \tau_{yy}(0) \qquad (7.3.34)$$

Here $p\ (0)$ and $\tau_{yy}\ (0)$ are values of the appropriate variables on a plane $y = 0$. The substitution of expressions (7.3.31) and (7.3.32) in equations (7.3.5) results in the equations, which coincide at $\gamma = 0$ and represent the definition of the angle $\theta(y)$:

$$\theta'' \sin\theta \cos\theta + (1 - 3\sin^2\theta)\,\theta'^2 = 0 \qquad (7.3.35)$$

According to [293], near the solid wall the eddies are extended along the flow, owing to the condition for the angle θ at plane $y = 0$ being formulated as

$$\sin\,\theta(0) = 0 \qquad (7.3.36)$$

Except for the trivial solutions $\theta = 0$ and $\theta = \pi$, which satisfy to boundary condition (7.3.36), equation (7.3.35) has the first integral

$$\theta'^2 \sin^2\theta \cos^4\theta = b^2 \qquad (7.3.37)$$

Here constant b is determined by the equality

$$b^2 = \theta_0'^2 \sin^2\theta_0 \cos^4\theta_0 \qquad (7.3.38)$$

Here θ_0 and θ'_0 are the angle θ and its derivative at the upper boundary $y = \Delta$ of the wall eddies zone, which is being modeled.

Let us rewrite equation (7.3.37) as

$$\theta' \sin\theta \cos^2\theta = \pm b, \qquad b > 0. \qquad (7.3.39)$$

Resolving it with boundary conditions (7.3.36), we obtain two solutions:

$$\cos\theta = \pm(1 - 3by)^{1/3}, \qquad (7.3.40)$$

$$\cos\theta = \pm(1 + 3by)^{1/3}. \tag{7.3.41}$$

In a physical sense, the range of solution (7.3.40) is defined by the inequality $0 < y < \Delta$. Solution (7.3.41) at positive b and y, at first sight, has no physical sense, but we keep it for further discussion, considering now (7.3.40). Substituting it in equation (7.3.33) and then integrating, we obtain the required general solution

$$\frac{U}{U_*} = \frac{2}{(2q^2 - 1)\kappa}\left(\sqrt{q^2 - 1}\,\text{arctg}\,\frac{t}{\sqrt{q^2 - 1}} + \frac{q}{2}\ln\frac{q - t}{q + t} \right) + C_1, \tag{7.3.42}$$

$$t = (1 - 3by)^{1/3},$$

$$\alpha = \frac{\mu_4}{2\,\mu_1}, \qquad 2q^2 = 1 + \sqrt{1 + 4\alpha},$$

$$q > 0, \qquad U_* = \sqrt{\frac{\tau_w}{\rho}}, \qquad \kappa = \frac{2\mu_1 b}{\rho U_*} \tag{7.3.43}$$

Here C_1 is a constant of integration.

Use of boundary conditions at plane $y = 0$ allows us to obtain specific solutions. If the ground surface is *smooth*, due to neglecting of the wall's laminar sublayer thickness and of velocity changes in it, the boundary condition will be

$$U_{|y=0} = 0 \tag{7.3.44}$$

Then constant C_1 is determined by the formula

$$C_1 = -\frac{2}{(2q^2 - 1)\kappa}\left(\sqrt{q^2 - 1}\,\text{arctg}\frac{1}{\sqrt{q^2 - 1}} + \frac{q}{2}\ln\frac{q - 1}{q + 1} \right) \tag{7.3.45}$$

Let us find a form of solution (7.3.42), (7.3.45) in the case when $3by << 1$. The analysis of experimental data [48] on the basis of the model considered here had shown [16] that the value q differs negligibly from unity.

Therefore, assuming that $q = 1$ in formulas (7.3.42), (7.3.45) always, when such substitution keeps the expression's sense, we obtain

$$\frac{U}{U_*} = \frac{1}{\kappa}\left[\ln\frac{1-(1-3by)^{1/3}}{1+(1-3by)^{1/3}} + \ln\frac{2}{q-1}\right]. \tag{7.3.46}$$

If numerator and denominator of the ratio in the first term of (7.3.46) are represented as linear functions of by, we obtain:

$$\frac{U}{U_*} = \frac{1}{\kappa}\left(\ln\frac{by}{2-by} + \ln\frac{2}{q-1}\right). \tag{7.3.47}$$

After simple transformations we have

$$\frac{U}{U_*} = \frac{1}{\kappa}\ln y_+ + C + \frac{b}{2\kappa}y \tag{7.3.48}$$

$$y_+ = \frac{U_* y}{\nu},$$

$$C = \frac{1}{\kappa}\ln\frac{b\nu}{(q-1)U_*},$$

Here ν is the kinematics' fluid viscosity. If we neglect the last term in the first formula of (7.3.48), the logarithmic wall velocity profile is

$$\frac{U}{U_*} = \frac{1}{\kappa}\ln y_+ + C, \tag{7.3.49}$$

Thus κ should be considered as the Karman constant. As $b > 0$, formula (7.3.48) corresponds to a steady stratification of an atmosphere; the formula (7.3.49) - to a neutral stratification [194, 195, 271]. The formulas, describing a structure of a wind at unstable stratification, will be considered below.

If the *ground surface is rough*, the roughness is characterized by a parameter h_0 such that [194, 195, 271] the boundary condition for velocity U has the form

$$U_{|y=h_0} = 0. \tag{7.3.50}$$

Then constant C_1 is determined by the formula

$$C_1 = -\frac{2}{(2q^2-1)\kappa}\left(\sqrt{q^2-1}\,\mathrm{arctg}\,\frac{t_0}{\sqrt{q^2-1}} + \frac{q}{2}\ln\frac{q-t_0}{q+t_0}\right),$$

$$t_0 = (1-3bh_0)^{1/3}. \tag{7.3.51}$$

Repeating the calculations used in the analysis of current at a smooth surface, at $q = 1$ and $3by << 1$ in the formulas (7.3.42) and (7.3.51) we get

$$\frac{U}{U_*} = \frac{1}{\kappa}\left[\ln\frac{y}{h_0} + \frac{b}{2}(y-h_0)\right], \quad h_0 \le y \le \Delta \tag{7.3.52}$$

Formula (7.3.52) coincides with the Obukhov-Monin solution [194, 195] for the steadily stratified atmosphere, as $b > 0$, although the sense of the cofactor of $(y - h_0)$ is another[24]. If the second term in formula (7.3.52) is neglected, the latter passes into the formula for wind structure in the neutrally stratified atmosphere [194, 195].

Let the boundary condition - now for the angle θ - be formulated again not at the plane $y = 0$ but at $y = y_1 > 0$, and looks like

$$\sin\,\theta(y_1) = 0. \tag{7.3.53}$$

Solving equation (7.3.39) with boundary conditions (7.3.53), we obtain two solutions:

[24] This formula, describing experimentally measured wind profiles, was obtained in [194] for the atmosphere stratification created by heat exchange. Here it is motivated by the effect of wall eddy mesostructure.

$$\cos\theta = \pm[1 - 3b(y - y_1)]^{1/3}, \qquad y_1 \le y \le (y_1 + \Delta) \tag{7.3.54}$$

$$\cos\theta = \pm[1 + 3b(y - y_1)]^{1/3}, \qquad y_1 \le y \le (y_1 + \Delta) \tag{7.3.55}$$

As above, we shall be limited, for a while, by (7.3.54). Substituting it into equation (7.3.33) and then integrating of the latter with the boundary condition

$$U_{|y=y_1+h_0} = 0 \tag{7.3.56}$$

we obtain the solution, which is determined by (7.3.42) and (7.3.51) provided that

$$t = [1 - 3b(y - y_1)]^{1/3}, \qquad t_0 = (1 - 3bh_0)^{1/3} \tag{7.3.57}$$

Repeating the previous analysis of this section at $q = 1$ and $3b\,(y - y_1) << 1$, from the formulas (7.3.42), (7.3.51) and (7.3.57) we get

$$\frac{U}{U_*} = \frac{1}{\kappa}\left[\ln\frac{y - y_1}{h_0} + \frac{b}{2}(y - y_1 - h_0)\right], \tag{7.3.58}$$

$$(y_1 + h_0) \le y \le (y_1 + \Delta)$$

If we neglect the second term in formula (7.3.58), the known formula will turn out to be [3, 4, 7]

$$\frac{U}{U_*} = \frac{1}{\kappa}\ln\frac{y - y_1}{h_0} \tag{7.3.59}$$

Parameter y_1 is named as *the replacement height* [194, 195] or *the displacement of a zero plane* [271]. The formula (7.3.59) can be used for a steadily stratified atmosphere.

Alternative general and special solutions

Let us consider formally equation (7.3.33) in the case, when the angle θ is defined by formula (7.3.41), although it is clear that at the real values of θ it is meaningless. We accept $\mu_1 < 0$ but keep $\mu_4 > 0$. Then after substitution of (7.3.41) both parts of equation (7.3.33) remain positive and it can be written as

$$dU = -\frac{\tau_w}{\mu_1 b}\frac{t^2 dt}{t^4 - t^2 + \alpha} \qquad (7.3.60)$$

$$t = (1 + 3by)^{1/3}, \qquad \alpha = -\frac{\mu_4}{2\mu_1}$$

Integrating expression (7.3.60), we get

$$\frac{U}{U_*} = \frac{1}{(2q^2 - 1)\kappa}\left(q \ln\frac{t - q}{t + q} - \sqrt{1 - q^2}\ln\frac{t - \sqrt{1 - q^2}}{t + \sqrt{1 - q^2}} \right) + C_2 \qquad (7.3.61)$$

$$\kappa = -\frac{2\mu_1 b}{\rho U_*}, \qquad 2q^2 = 1 + \sqrt{1 - 4\alpha}, \qquad q > 0$$

Here C_2 is an integration constant that will be determined by the boundary conditions.

Smooth surface

Use of boundary condition (7.3.44) gives

$$C_2 = -\frac{1}{(2q^2 - 1)\kappa}\left(q\ln\frac{1-q}{1+q} - \sqrt{1-q^2}\ln\frac{1-\sqrt{1-q^2}}{1+\sqrt{1-q^2}} \right) \qquad (7.3.62)$$

As above, let assume $q = 1$ in formulas (7.3.61) and (7.3.62) everywhere, where such replacement is permissible. Then

$$\frac{U}{U_*} = \frac{1}{\kappa}\left(\ln \frac{(1 + 3by)^{1/3} - 1}{(1 + 3by)^{1/3} + 1} - \ln \frac{1 - q}{2} \right). \tag{7.3.63}$$

If $3by << 1$, formula (7.3.63) will be transformed to the form

$$\frac{U}{U_*} = \frac{1}{\kappa} \ln y_+ + C - \frac{b}{2\kappa} y \tag{7.3.64}$$

$$C = \frac{1}{\kappa} \ln \frac{b\nu}{(1 - q)U_*}$$

As $b > 0$, formula (7.3.64) corresponds to a structure of a wind at unstable stratification [194, 195]. If we neglect the last term, the logarithmic wall profile follows.

Rough surface (1)

Using a boundary condition (7.3.50), we obtain

$$C_2 = -\frac{1}{(2q^2 - 1)\kappa}\left(q \ln \frac{t_0 - q}{t_0 + q} - \sqrt{1 - q^2} \ln \frac{t_0 - \sqrt{1 - q^2}}{t_0 + \sqrt{1 - q^2}} \right)$$

$$t_0 = (1 + 3bh_0)^{1/3}. \tag{7.3.65}$$

At $q = 1$ and $3by << 1$ in formulas (7.3.61) and (7.3.65), the formula follows

$$\frac{U}{U_*} = \frac{1}{\kappa}\left[\ln\frac{y}{h_0} - \frac{b}{2}(y - h_0)\right], \tag{7.3.66}$$

$$h_0 \leq y \leq \Delta,$$

It coincides with the Obukhov-Monin formula for unstable stratification.

Rough surface (2)

Substitution of (7.3.55) in equation (7.3.33) under condition $\mu_l < 0$ and integrating the latter with boundary condition (7.3.56) gives the required solution as formulas (7.3.61) and (7.3.65), in which it is necessary to put

$$t = [1 + 3b(y - y_1)]^{1/3}, \quad t_0 = (1 + 3bh_0)^{1/3} \quad . \tag{7.3.67}$$

At $q = 1$ and $3by << 1$ this solution describes a wind velocity profile in unstable atmosphere stratification, close to a very rough surface

$$\frac{U}{U_*} = \frac{1}{\kappa}\left[\ln\frac{y - y_1}{h_0} - \frac{b}{2}(y - y_1 - h_0)\right], \tag{7.3.68}$$

$$(y_1 + h_0) \leq y \leq (y_1 + \Delta)$$

If we neglect the second term in brackets, this result turns into the known formula (7.3.59).

Thus, the formulas (7.3.41) and (7.3.55) result in solutions that have physical sense. As to formulas (7.3.41) and (7.3.55), they are completely clear, if angle θ is considered as an imaginary number, that is, to put $\theta = i\varphi$, where i - imaginary unit and φ - a real number. Then

$$\cos\theta = \mathrm{ch}\,\varphi \geq 1.$$

However in this case $\sin\theta = i\,\mathrm{sh}\,\varphi$ and the vector - director ℓ_i should be considered as a vector with complex coordinates. Nevertheless, it is simple to check, that at $\theta = i\varphi$ formula (7.3.38) is valid with real b.

As it is seen, in the frame of the considered model, the known logarithmic wind profiles in an atmosphere ground layer correspond to the approximate special solutions of the model equations, adequate to various boundary conditions for fluid velocity and an inclination angle of a vector-director. The empirical constants of classical profiles (the Karman constant, etc) are expressed through the considered model parameters.

7.4. FLOWS BETWEEN PLATES

Couette flow

Let a fluid flow between infinite parallel planes: one of the planes is motionless, and the other is moving with constant velocity U_0. For our system of coordinates, the x -axis coincides with the immovable plane and is directed as the flow; the y -axis is perpendicular to both planes and is directed to the moving plane. The immovable plane is $y = 0$ and the moving plane is above it at $y = 2H$.

Here we shall use[25] equation system (7.3.4) - (7.3.5) with its integrals (7.3.6) and (7.3.9). The simplification, shown in Section 7.3, permits us to use expressions (7.3.29) - (7.3.32).

Again, we obtain (7.3.33), (7.3.34), that is,

$$(\mu_1 \sin^2\theta \cos^2\theta + \frac{\mu_4}{2})\, U' = \tau_w \qquad (7.4.1)$$

$$p(y) = p(0) + \tau_{yy}(y) - \tau_{yy}(0) \qquad (7.4.2)$$

However, now $p\,(0)$, $\tau_{yy}\,(0)$ is the value of the appropriate variable at the bottom plane $y = 0$.

The equation for the angle θ has the same form as (7.3.35):

---

[25] Babkin V.A. Plane turbulent Couette flow (in press).

$$\theta'' \sin\theta \cos\theta + \theta'^2 (1 - 3\sin^2\theta) = 0. \tag{7.4.3}$$

Assuming that the immovable plane is smooth, the boundary conditions at the walls will be formulated as

$$\begin{aligned} &U(0) = 0, \qquad U(2H) = U_0 \\ &\sin\theta(0) = 0, \quad \sin\theta(2H) = 0 \end{aligned} \tag{7.4.4}$$

The second boundary condition of (7.4.4) reflects the fact that turbulent eddies are extended along the flow at a solid wall [39, 191, 233].

The first integral of (7.4.3)

$$\theta'^2 \sin^2\theta \cos^4\theta = b^2, \tag{7.4.5}$$

includes the constant b determined by the equality

$$b^2 = \theta_0'^2 \sin^2\theta_0 \cos^4\theta_0 \tag{7.4.6}$$

Here θ_0 and θ'_0- angle θ and its derivative at the outer boundaries of the vortex sublayers attached to the walls. The same value b^2 is accepted at both plane sublayers.

The solution of equation (7.4.5) with the second boundary condition (7.4.4) looks like

$$\cos\theta = \pm(1 - 3by)^{1/3}, \qquad 0 \le y \le y_1 \tag{7.4.7}$$

$$\cos\theta = \mp[1 - 3b(2H - y)]^{1/3}, \qquad y_2 \le y \le 2H \tag{7.4.8}$$

Here y_1 and y_2 are coordinates where condition (7.4.6) is valid.

The signs combination (±) corresponds to $\theta = 0$ at the immovable plane and $\theta = \pi$ at the movable one. The change of signs corresponds to the contrary order of angles.

Correspondingly, we obtain from (7.3.29) - (7.3.32) the velocity profiles as

$$U(y) = \begin{cases} \Psi(t) & t = (1 - 3b\,y)^{1/3}, \quad 0 \le y \le y_1 \\ U_0 - \Psi(t) & t = [1 - 3b(2H - y)]^{1/3}, \quad (2H - y_2) \le y \le 2H \end{cases} \tag{7.4.9}$$

$$\Psi(t) = \frac{2U_*}{\kappa(2q^2 - 1)} \left[\sqrt{q^2 - 1}\left(\operatorname{arctg}\frac{t}{\sqrt{q^2 - 1}} + \operatorname{arctg}\frac{1}{\sqrt{q^2 - 1}} \right) + \frac{q}{2}\left(\ln\frac{q - t}{q + t} - \ln\frac{q - 1}{q + 1} \right) \right] \tag{7.4.10}$$

$$\alpha = \frac{\mu_4}{2\,\mu_1}, \quad 2q^2 = 1 + \sqrt{1 + 4\alpha}, \quad q > 0,$$

$$U_* = \sqrt{\frac{\tau_w}{\rho}}, \quad \kappa = \frac{2\mu_1 b}{\rho u_*} \tag{7.4.11}$$

For comparison of formula (7.4.9) with test data we need to determine constant b.

If the sublayers are touching each other and because the velocity profile is smooth, one can find this constant from the equation:

$$U = 2\Psi(t_H), \qquad t_H = (1 - 3bH)^{1/3} \tag{7.4.12}$$

In Figure 7.4.1 the experimental points [74] are shown as well as diagrams of three velocity profiles U/U_* as a function of $y_+ = (U_* y)/\nu$ in the zone $0 \le y \le y_1$. Curve *1* is calculated by (7.4.9) for the conditions of these experiments. Here $\nu = 1.486 \cdot 10^{-5}\ m^2/s$ is kinematical fluid (air) viscosity, $\rho = 1.21\ kg/m^3$ is its density.

Parameters, varying in experiments, and also the model parameters, necessary for calculations, are given in Table 7.4.1 where the following designations are used: U_0 - velocity of the mobile plane, H - half of the distance between planes. At $y_+ >$

20 all experimental points are near curve *1*, which represents three practically coinciding curves (7.4.9), calculated for the variants of Table 7.4.1. At $y_+ < 20$ there is a laminar boundary layer that explains the deviation of the calculated curves and the experiment.

Table 7.4.1.

Test	U_0, *m/s*	*H*, *mm*	U_*, *m/s*	*b*, *1/m*	μ_1, *kg/(m s)*	μ_4, *kg/(m s)*	*q*
I	12.84	22	0.293	3.44	0.017	$3.81\cdot 10^{-6}$	1.000056
II	12.84	33	0.282	3.37	0.017	$3.88\cdot 10^{-6}$	1.000057
III	17.08	33	0.363	3.42	0.021	$3.78\cdot 10^{-6}$	1.000045

Here the turbulent viscosity is identified with the following parameter:

$$\nu_t = \frac{\mu_1}{\rho}\sin^2\theta\cos^2\theta \tag{7.4.14}$$

Let us show, that near the solid wall, outside of the laminar sublayer, according to formula (7.4.9), the logarithmic profile of velocity (curve 2) follows. As $q \approx 1$, assuming that $q = 1$ everywhere, where such replacement does not lose its sense, we get

$$\frac{U}{U_*} = \frac{1}{\kappa}\left(\ln\frac{1-t}{1+t} - \ln\frac{q-1}{2}\right). \tag{7.4.15}$$

Let $|3by| << 1$. Then

$$t = 1 - by, \quad \ln(1+t) = \ln 2 - \frac{by}{2}. \tag{7.4.16}$$

Substituting these expressions into formula (7.4.15) and neglecting term $(by/2)$ in comparison with $\ln 2$, we get

$$\frac{U}{U_*} = \frac{1}{\kappa}\left(\ln y + \ln\frac{b}{q-1}\right). \tag{7.4.17}$$

This formula can be rewritten in the known form

$$\frac{U}{U_*} = \frac{1}{\kappa}\ln\frac{U_* y}{\nu} + C, \qquad C = \frac{1}{\kappa}\ln\frac{b\nu}{(q-1)U_*}. \tag{7.4.18}$$

In Figure 7.4.1 the graph, corresponding to (7.4.9) at $\kappa = 0.334$ and $C = 3.40$, is given by curve *2*. In the interval $5 \le y_+ \le 300$ curves (7.4.14) and (7.4.18) differ no more than by *1 %*, but also at bigger y_+ the distinction is insignificant. Though the values κ and C obviously depend on parameters of model and flow, nevertheless, it appears that there is an interval of y_+ where they are constant. Formulas (7.4.10) and the second formula of (7.4.15) show the performance of conditions necessary for the constancy κ and C at different velocities

$$\mu_4 = const, \qquad \frac{\mu_1 b}{U_*} = const.$$

According to Table 7.4.1 in experiments [74] these conditions are really carried out.

The authors of [74] used formula (7.4.18) at $\kappa = 0.39$ and $C = 5.2$ for analytical approximation of the experimental results, which diagram in Figure 7.4.1 is given by curve *3*. However, obviously, this curve approximates the worst test points in comparison with curve 2.

Using the boundary conditions, similar to (7.4.4), in the mobile plane results in a solution for the wall vicinity in this plane. If the distance between planes is big enough, the zones close to the wall can overlap and to get the solution between them it is necessary to use another model as was done for example in [16].

Let us find the stress τ_w that acts on the unit of the moving plate to provide its velocity U_0. In the case when wall zones are touching at the middle plane, that is, at $y_1 = y_2 = H$, then this value has to be found from solution (7.4.9) - (7.4.11).

To avoid huge formulas, let us assume that $q = 1$ always, when it is permissible. Then, recalling that $u(H) = U_c / 2$ and using the expression U_*, we obtain

$$\tau_w = \frac{\mu_1 b U_c}{L},$$

$$L = \ln\frac{1-(1-3bH)^{1/3}}{1+(1-3bH)^{1/3}} + \ln\frac{2}{q-1}. \tag{7.4.19}$$

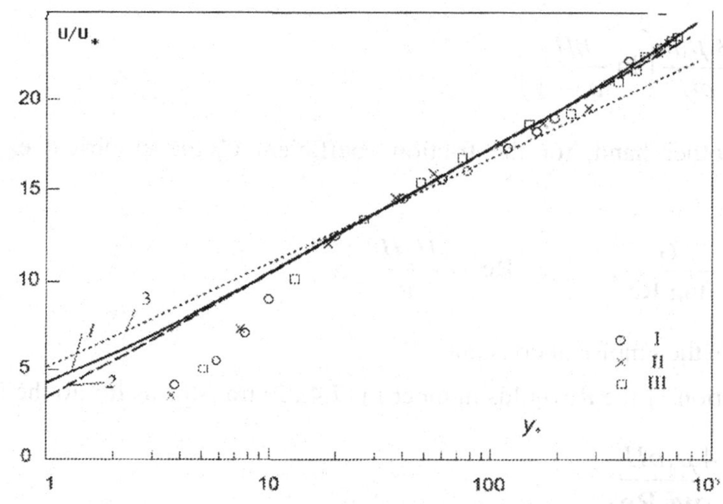

Figure 7.4.1. Scaled velocity profile in comparison with test data

It could appear that τ_w is a linear function of U_0. However, it is not so, because parameters μ_1, b and q depend on the mesovortex structure of a flow, that, in its turn, is a functional of the velocity U.

Let us find the friction coefficient C_f on the basis of previous solution for the Couette flow and compare it with experimental data [74]. Then we get:

$$C_f = \frac{2\tau_w}{\rho U_c^2}, \qquad U_c = \frac{U}{2}, \tag{7.4.20}$$

Here U_c is the fluid velocity at the middle plane $(y = H)$ and U_0 is the moving plane velocity.

Expressions (7.4.19) - (7.4.20) give the following result:

$$C_f = \frac{8\mu_1 b}{\rho L U}. \tag{7.4.21}$$

If $bh << 1$, formula (7.4.21) has the form

$$C_f = \frac{8\mu_1 b}{\rho U}\left(\ln\frac{bH}{q-1}\right)^{-1}. \tag{7.4.22}$$

On the other hand, for the friction coefficient C_f the empirical expression is known:

$$\sqrt{\frac{C_f}{2}} = \frac{G}{\log \mathrm{Re}}, \qquad \mathrm{Re} = \frac{U_c H}{\nu}, \tag{7.4.23}$$

Here G is the empirical constant.

Introduction of the Reynolds number in (7.4.22) transforms it into the following

$$C_f = \frac{4\mu_1 bH}{\rho\nu L \,\mathrm{Re}}. \tag{7.4.24}$$

Comparing values C_{f1}, determined by (7.4.23) at $G = 0.182$ [74], with the values C_{f2}, calculated by (7.4.24) with usage of Table 7.4.1, we have for experiments I - III, correspondingly: $1000\, C_{f1} = 4.19; 3.84; 3.62$ and $1000\, C_{f2} = 4.23; 3.84; 3.55$. One can see that results are practically identical. Expression (7.4.22) also gives very close values.

Head flow between parallel planes

Consider incompressible fluid flow between two infinite parallel planes under action of constant pressure drop [14, 16]. The system of Cartesian coordinates x, y, z is such that axis x is along the flow, axis y is perpendicular to the planes, axis z forms right hand trial axes x, y. The equations of the planes are $y = \pm H$.

Instead of (7.4.1) - (7.4.2) the following system for U and θ will be used now:

$$(\mu_1 \sin^2\theta \cos^2\theta + \frac{\mu_4}{2})U' = -\tau_w \frac{y}{H}, \qquad \tau_w = -\frac{\partial p}{\partial x}H \tag{7.4.25}$$

$$\theta'' \sin\theta \cos\theta + \theta'^2(1 - 3\sin^2\theta) = 0 \tag{7.4.26}$$

The boundary conditions at smooth walls are:

$$\sin\theta(\pm H) = 0, \qquad U(\pm H) = 0. \tag{7.4.27}$$

Almost repeating calculation of the previous problem, we shall get the solution of the system (7.4.25) and (7.4.26) as

$$t = [1 - 3b(H - |y|)]^{1/3} \tag{7.4.28}$$

$$\frac{3\mu_1 b^2 H}{\sqrt{\rho\tau_w}}\frac{U}{U_*} = \frac{3bH - 1}{2q^2 - 1}\left[\begin{matrix}\sqrt{q^2 - 1}\,\mathrm{arctg}\dfrac{t}{\sqrt{q^2 - 1}} + \\ \dfrac{q}{2}\ln\dfrac{q - t}{q + t}\end{matrix}\right] + \frac{t^2}{2} +$$

$$+\frac{1 + 2\alpha}{4(2q^2 - 1)}\ln\frac{q^2 - t^2}{t^2 + q^2 - 1} + \frac{1}{4}\ln\left|t^4 - t^2 - \alpha\right| + C_0, \tag{7.4.29}$$

$$C_0 = \frac{1 - 3bH}{2q^2 - 1}\left[\sqrt{q^2 - 1}\,\mathrm{arctg}\frac{1}{\sqrt{q^2 - 1}} + \frac{q}{2}\ln\frac{q - 1}{q + 1}\right] -$$

$$\frac{1 + 2\alpha}{4(2q^2 - 1)}\ln\left(1 - \frac{1}{q^2}\right) - \frac{\ln\alpha}{4} - \frac{1}{2}, \tag{7.4.30}$$

Here α, q, U_* are values, determined by (7.4.11).

In Figure 7.4.2 the experimental points [48] are given as well as theoretical curves (7.4.28) - (7.4.30) for the corresponding conditions. Air was used as a fluid (density $\rho = 1.205\ kg/m^3$, kinematical viscosity $\nu = 1{,}500\ 10^{-5}\ m^2/s$). Distance between planes $2H = 0.18\ m$. Reynolds number is $Re = (\langle U\rangle H/\nu) =$ *57,000; 120,000; 230,000,* where $\langle U\rangle$ is the average flow velocity. The parameter values used in calculations are given in Table 7.4.2.

The graphs show that solution (7.4.28) - (7.4.30) is not worse compared to the experimental data than the universal logarithmic law [48, 110] in the interval $y_+ \geq 20$, where $y_+ = (H - |y|)u_*/\nu$ is the dimensionless distance from the wall, whereas at the wall vicinity theory and experiments are essentially different.

Table 7.4.2

Re	U_*, m/s	b, 1/m	μ_1, kg/m·s	μ_4, kg/m s	q
57 000	0.39	4.10	0.021	1.43 10^{-6}	1.000017
120 000	0.80	3.81	0.047	2.44 10^{-6}	1.000013
230 000	1.36	4.12	0.073	1.49 10^{-6}	1.0000051

At $y_+ < 20$ the turbulence of the flow is considerably weakened and for its adequate description other models are required, for example, models of viscous laminar and transitive sublayers [110].

We shall show that "the logarithmic law of a wall" follows [48, 110] from (7.5.4) - (7.5.6) under certain conditions. Due to the flow symmetry relative to the plane $y = 0$, we shall be limited to values $y \geq 0$. Assuming, as above, that $q = 1$ everywhere in the solution, where such replacement does not lose the sense, and fulfilling the elementary transformations, we get

$$\frac{3\mu_1 b^2 H}{\sqrt{\rho\tau_w}}\frac{U}{U_*} = \frac{3bH}{2}\ln\frac{2(1-t)}{q-1} + \left(1 - \frac{3bH}{2}\right)\ln(1+t) + \frac{t^2-1}{2} - \ln 2 \tag{7.4.31}$$

For $3b(H - y) << 1$, we have

$$1 - t = b(H - y),$$
$$1 + t = 2 - b(H - y) \approx 2, \tag{7.4.32}$$
$$t^2 = 1 - 2b(H - y) \approx 1$$

The substitution of (7.4.32) into formula (7.4.31) gives

$$\frac{\mu_1 b}{\sqrt{\rho\tau_w}}\frac{U}{U_*}=\frac{1}{2}\ln\frac{b(H-y)}{q-1} \tag{7.4.33}$$

This formula will be easily transformed to the standard form

$$\frac{U}{U_*}=\frac{1}{\kappa}\ln\frac{(H-y)\,U_*}{\nu}+C \tag{7.4.34}$$

$$\kappa=\frac{2\mu_1 b}{\sqrt{\rho\tau_w}}=\frac{2\mu_1 b}{\rho U_*}, \qquad C=\frac{1}{\kappa}\ln\frac{b\nu}{(q-1)U_*}.$$

In Table 7.4.3 the values κ and C, calculated by formulas (7.4.33) with use of Table 7.4.2 data (columns 2 and 4), are given as well as the test values of these variables [48] (they are κ_e and C_e, columns 3 and 5).

Now we shall express the flow rate Q in the channel with pressure drop $\Delta p = p_1 - p_2$, in corresponding cross-sections x_1 and $x_2 = x_1 + l$.

$$Q=2\int_0^h U\,dy=2H<U>, \tag{7.4.35}$$

Hydraulic head h for the head flow between two parallel planes, separated by distance $2H$ is determined as

$$h=\phi\frac{<U>^2}{2D}, \qquad \phi=\frac{8\tau_w}{\rho<U>^2}, \qquad D=4H \tag{7.4.36}$$

Here ϕ is the resistance coefficient, D is the hydraulic diameter, $<U>$ is the average flow velocity and g is the gravity acceleration.

As one can see in the last column of Table 7.4.3, expression (7.4.34) determines the velocity profile in a wall zone that does not coincide with the flow interval. Moreover, it loses its sense out of the zone and cannot be extrapolated.

However, approximately (7.4.34) can be used in the entire interval: $0 \le y < H$. Assuming, as in [30], that this velocity profile is valid everywhere in $0 \le y < H$, and substituting it into (7.4.35), just changing κ and C to their expression (7.4.28), we obtain

$$<U> = -\frac{\rho U_*^2}{2\mu_1 b}\left(1 - \ln\frac{bH}{q-1}\right). \tag{7.4.37}$$

Because

$$\tau_w = \rho U_*^2 = \frac{\Delta p}{l} H \tag{7.4.38}$$

substitution of (7.4.37) into (7.4.35) yields

$$Q = -\frac{\Delta p}{l}\frac{H^2}{\mu_1 b}\left(1 - \ln\frac{bH}{q-1}\right) \tag{7.4.39}$$

As well as in the Couette flow, dependency of Q on $\Delta p/l$ is not linear, because parameters μ_1, b and q are functions of the global flow parameters – flow rate and pressure drop. Introducing the Reynolds number $\mathrm{Re} = <U> H/\nu$, we obtain

$$\phi = -\frac{16\mu_1 bH}{\rho\nu \mathrm{Re}}\left(1 - \ln\frac{bH}{q-1}\right)^{-1}. \tag{7.4.40}$$

Therefore, the resistance coefficient φ is a function of dimensionless combinations *Re*, *bh*, $(\mu_1/\rho\nu)$ and *q*.

It is easy to see that the calculated and experimental values κ and C are close enough. Besides κ и C are determined here by the same formulas, as in the Couette flow, and because the flow conditions are essentially different in these cases, then values κ и C differ also.

The extreme right points of curves *1 - 3* correspond to values t = 0. Formally they define the upper boundary of wall turbulent flow. However deviation of experimental points from the theory begins a little bit earlier, approximately at. $y_+ = 1200$.

It is natural to accept distance from the wall $(H - y)$, appropriating to this value y_+, as the thickness of wall turbulent layer δ. The ratio δ/h at various Reynolds numbers is given in Table 7.4.3. With the flow velocity increase, the layer thickness is decreasing, which quite corresponds to the analysis [249], according to which the maximal distance of system of Λ-eddies from the solid wall decreases with velocity growth.

Table 7.4.3

Re	κ	κ_e	C	C_e	δ/H
57 000	0.366	0.37	6.09	5.9	0.51
120 000	0.372	0.37	4.59	4.5	0.25
230 000	0.367	0.365	5.98	6.0	0.15

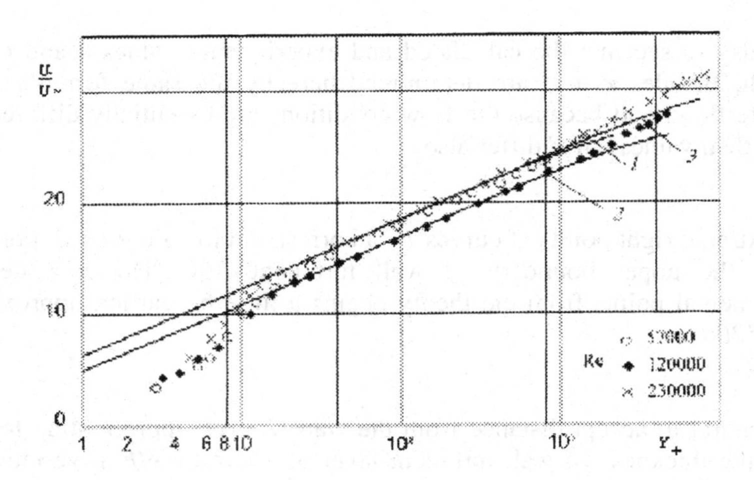

Figure 7.4.2. Calculations and experiment for head flow in a plane channel

8. LITERATURE

[1] Aero E.L., Bulygin A.N., Kuvshinsky E.V. *Asymmetric hydrodynamics.* J. Appl. Math. Mech. (PMM), v. 29, # 2 (1964).

[2] Aero A.L., Bulygin A.N. *Hydromechanics of liquid crystals.* Advances (Itogi) in Science and Technology: Hydromechanics, v.7, Moscow: VINITI (1973).

[3] Afanasiev E.F., Nikolaevskiy V.N. *On the development of asymmetrical suspension hydrodynamics with rotating solid particles.* In: Problems of Hydromechanics and Continuum Mechanics (the Sedov's 60-th Anniversary). SIAM, Philadelphia, 16 – 26 (1969).

[4] Ahmadi G., Koh S.L., Goldschmid V.M. *A theory of nonsimple microfluids.* In: Recent Advances in Engineering Science. V. 5, part 2. Gordon & Breach, New York, 9 – 20 (1970).

[5] Ahmadi G., Goldschmid V.M., Dean B. *A model of incompressible turbulent shear flow.* Iranian J. Sci. Technol., v. 5, # 4, 147 – 158 (1976).

[6] Allen S.J., Kline K.A., Ling C.C. *Transient shear flow of fluids with deformable microstructure.* Acta Mech., v.18, 1-20 (1973).

[7] Arsen'ev S.A., Nikolaevskiy V.N. *Vertical structure of oceanic currents at the Equator with regards for mesoscale vortices.* Doklady Earth Sciences (DAN), v. 377A, # 3, 365 – 367 (2001).

[8] Arseniev S.A., Nikolaevskiy V.N. *Turbulent-vortex flows in channels and heat-coolers of atomic energy stations.* Atomic Energy, v.90, # 5 (2001).

[9] Arseniev S.A., Nikolaevskiy V.N., Shelkovnikov N.K. *Tornado: origin, evolution and stability.* Vestnik MGU. Ser. 3: Physics and Astronomy, #1, 50 – 53 (2000).

[10] Artemov M.A., Nikolaevskiy V.N. *On equations of asymmetric turbulence in magneto-hydrodynamics.* Letters in Appl. Engng Sci., v. 4, # 3, (1976).

[11] Artemov M.A., Ivchenko V.O., Nikolaevskiy V.N. *On parameterization of mesoscale eddies in numerical models of large-scale oceanic circulation.* Proc. Arctic's and Antarctic's Institute. V. 387. Leningrad: Gidrometeoizdat (1982).

[12] Atkin R.J., Fox N. *A multipolar approach to liquid helium II.* Acta Mechanica, v.21, 221 – 239 (1975).

[13] Atkin R.J., Fox N. *Acoustic wave propagation in liquid helium II.* J. Sound & Vibration, v. 42, # 1, 13 – 29 (1975).

[14] Babkin V. A. *Anisotropic turbulence of incompressible fluid flow between plane parallel walls.* J. Appl. Math. Mech. (PMM), v.49, # 3 (1985).

[15] Babkin V. A. *Anisotropic turbulence at incompressible fluid flow between coaxial rotating cylinders.* J. Appl. Math. Mech. (PMM), v.52, # 2 (1988).

[16] Babkin V.A. *Turbulent stream in a wall zone as a flow of anisotropic fluid.* Eng.-Phys. J. (Minsk), v. 75, # 5, 69 – 73 (2002).

[17] Barenblatt G.I., Zeldovich Ya. B. *Self-similar solutions as intermediate asymptotics.* Annual Review of Fluid Mechanics. V. 4, 285 – 312 (1972).

[18] Bagriantzev N.V., Danilov A.I., Ivchenko V.O., Nikolaevsky V.N. *Orientation effects in geophysical fluid dynamics.* Int. J. Eng. Sci., v. 21, # 7 (1983).

[19] Batchelor G. K. *The theory of axisymmetric turbulence.* Proc. Roy. Soc. London, A186, 480 – 502 (1946).

[20] Batchelor G.K. *The Theory of Homogeneous Turbulence.* Cambridge University Press (1953).

[21] Batchelor G.K. *An Introduction to Fluid Dynamics.* Cambridge University Press. (1970).

[22] Batchelor G.K. *The stress system in a suspension of free force particles.* J. Fluid Mech., v. 41, pt 3, 545 – 570 (1970).

[23] Bear J. *Dynamics of Fluids in Porous Media.* New York: Elsevier (1972).

[24] Bengitson L., Lighthill J., Eds. *Intense Atmospheric Vortices.* Springer-Verlag, New York (1982).

[25] Beran M.J. *Statistical Continuum Theories.* New York: Interscience (1968).

[26] Beresnev I.A., Nikolaevskiy V.N. *A model for nonlinear seismic waves in a medium with instability.* Physics D, v. 66, 1 – 6 (1993).

[27] Berezin Yu. A., Trofimov V.M. *Thermal convection in a non-equilibrium turbulent medium with rotation.* Fluid Dynamics, v. 29, # 6 (1994).

[28] Berezin Yu. A., Trofimov V.M. *A model of non-equilibrium turbulence with an asymmetric stress. Application to the problems of thermal convection.* Continuum Mech. Thermodynamics, v. 7, 415 – 437 (1995).

[29] Berezin Yu. A., Trofimov V.M. *Large-scale vortex generation driven by non-equilibrium turbulence.* Fluid Dynamics. V. 31, # 1 (1996).

[30] Birch S.F., Lebedev A.B., Lyubimov D.A., Secundov A.N. *Modeling of turbulent 3D jet and boundary layer streams.* Fluid Dynamics, # 5 (2001).

[31] Birkhoff G., Zarantanello E.H. *Jets, Wakes and Cavities.* New York: Acad. Press, (1957).

[32] Blackwelder R.F., Kovasznay L.S.G. *Time scales and correlations in a turbulent boundary layer.* Phys. Fluids, v.9, # 9 (1972).

[33] Brenner H. *Rheology of two-phase systems.* Annual Review Fluid Mechanics, v. 2, 137 – 176 (1975).

[34] Brekhovskikh L.M. et al. *Some results of hydrophysical experiment at ground of Tropical Atlantics.* Trans. USSR Acad. Sci. (Izvestia) Physics of Atmosphere and Ocean, v.7, # 5 (1971).

[35] Bubnov V.A., Martynenko O.G., Kaliletz V.I., Ustok H.Z. *Radial force of isolated vortex.* Minsk, Inst. Heat - Mass Exchange, 21 p. (1987).

[36] Buevich Y.A., Nikolaevskiy V.N. *Equations for moments of averaged turbulence with anisotropy of vortex type.* Proc. USSR Acad. Sci. (DAN), v. 201, # 2, 288 – 291 (1971).

[37] Buevich Yu. A., Nikolaevskiy V.N. *Theory of turbulence with anisotropy of eddy type.* In: Continuum Mechanics and Related Problems of Analysis. (The N.I. Muskhelishvili's 80th anniversary). Moscow, Nauka (1972).

[38] Busse F. H. *Magnetohydrodynamics of the Earth's dynamo.* Annual Review of Fluid Mechanics. V. 10, 435 – 462 (1978).

[39] Cantwell B. J. *Organized motion in turbulent flow.* Annual Review Fluid Mechanics, v. 13, 457 – 515 (1981).

[40] Chandrasekhar S. *The theory of axisymmetric turbulence.* Phil. Trans. Roy. Soc. London, A242, 557 – 577 (1950).

[41] Chandrasekhar S. *The decay of axisymmetric turbulence.* Proc. Roy. Soc. London, A203, 358 – 364 (1950).

[42] Chandrasekhar S. *Liquid Crystals.* Cambridge University Press (1977).

[43] Charney J.G. *Geostrophic turbulence.* J. Atmos. Sci., v. 28, 1087 – 1095 (1971).

[44] Charney J.G. *Nonlinear theory of a wind driven homogeneous layer near the Equator.* Deep-Sea Res., v.6, # 4, 303 – 310 (1960).

[45] Chevray R. *The turbulent wake of a body of revolution.* Trans. Amer. Mech. Eng., ser D (J. Basic Engng.), v.90, # 2 (1968).

[46] Chevray R., Kovasznay L.G.S. *Turbulence measurements in the wake of a thin plate.* AIAA J., v.7, # 8 (1969).

[47] Chimonas G., Hauser H.M. *The transfer of angular momentum from vortices to gravity swirl waves.* J. Atmos. Sci., v. 54, 1701 - 1711 (1997).

[48] Comte-Bellot G. *Ecoulement Turbulent Entre Deux Parois Paralleles.* Edite Scient. Techn. Paris (1965).

[49] Condiff D.W., Brenner H. *Transport mechanics in the system of orientable particles.* Physics of Fluids, v. 12, # 3, 539 – 551 (1969).

[50] Condiff D.W., Dahler J.S. *Fluid mechanical aspects of antisymmetric stress.* Physics of Fluids, v.7, # 6, 842 – 854 (1964).

[51] Corino E.R., Brodkey R.S. *A visual investigation of the wall region in turbulent flow.* J. Fluid Mech., v. 37, # 1, 1-30 (1969).

[52] Corrsin S. *Limitations of gradient transport models in random walks and in turbulence.* Advances in Geophysics. V. 18A, 25 – 60 (1974).

[53] Corrsin S., Kistler A. *Free-steam boundaries of turbulent flows.* NASA Report 1244 (1965).

[54] Cosserat E. et F. *Theorie des Corps Deformables.* Paris: Herman (1909).

[55] Cox J. P. *Theory of Stellar Pulsation.* Princeton University Press (1980).

[56] Dahler J.S. *Transport phenomena in a fluid composed of diatomic molecules.* J. Chem. Phys., v. 30, p. 1447 (1959).

[57] Dahler J.S., Scriven L. *Angular momentum of continua.* Nature, # 4797, Oct. 7 (1961).

[58] Dahler J.S., Scriven L. *Theory of structured continua: general consideration of angular momentum and polarization.* Proc. Roy. Soc., A275, 504 – 527 (1963).

[59] Daily J.W. *Some aspects of flowing suspensions. Proc. IX Midwestern Mechanics* Conference. Madison, Wisconsin., August 16-18 (1965).

[60] Daily J.W., Roberts P.R. *Rigid particle suspensions in turbulent shear flow: some size effects with spherical particles.* Tappi, v. 49, # 3, 115 – 125 (1966).

[61] Danilov A.I., Ivchenko V.O., Nikolaevskiy V.N. *On large-scale circulation of barotropic ocean with parameterization of synoptic eddies.* Proc. USSR Acad. of Sci. (DAN), v.262, # 6 (1982).

[62] Danilov A.I. *Non-stationary barotropic circulation with account of action of sub-net scale.* Trans. of Arctic and Antarctic Institute, v. 387. Leningrad: Gidrometeoizdat (1982).

[63] De Groot S.R., Mazur P. *Non-equilibrium Thermodynamics.* Amsterdam: North-Holland (1962).

[64] De Groot S.R., Sattorp L.H. *Electrodynamics.* Amsterdam, North-Holland (1972).

[65] De Gennes P. G. *The Physics of Liquid Crystals.* Oxford, Clarendon (1974).

[66] Dopazo C. *On conditioned averages for intermittent turbulent flows.* J. Fluid Mechanics, v.81, pt 3 (1977).

[67] Dryden H.L. *A review of the statistical theory of turbulence* (1942). In: "Classic Papers on Statistical Theory", Friedlander S. K., Topper L., Eds. New York: Interscience, 115 – 150 (1961).

[68] Dryden H.L., Murnaghan F.D., Bateman H. *Hydrodynamics.* Dover Publications Inc., New York (1956).

[69] Dryden W.A. *Effects on the scale of spatial averaging on the kinetic energies of small – scale turbulent motion.* J. Meteorology, v. 14, # 4, 287 – 292 (1957).

[70] Eckman J.P. Jarvenpaa E., Jarvenpaa M. *Porosities and dimensions of measures.* Nonlinearities, v. 13, 1 – 18 (2000).

[71] Egger J. *Angular momentum of β - plane flows.* J. Atmos. Sci., v. 58, 2502 – 2508 (2001).

[72] Egger J., Hoinka K.-P. *Mountain torques and the equatorial components of global angular momentum.* J. Atmos. Sci., v. 57, 2319 – 2331 (2000).

[73] Einstein A. *Eine neue bestimmung der molekuldimensionen.* Annalen der Physik, v. 19, 289 – 306 (1906).

[74] El Telbany M. M. M. and Reynolds A.J. *The structure of turbulent plane Couette flow.* Transactions of ASME. Journal of Fluids Engineering, v. 104, # 3, 367 – 372 (1982).

[75] Emanuel K.A. *The theory of hurricanes.* Ann. Rev. Fluid Mech., v. 23, 179 – 196 (1991).

[76] Erechnev D.A. Leontiev D.I., Melnikova O.N. *Generation of cylindrical vortexes at a bottom that resists to water flows.* Physics of Atmosphere and Ocean, v. 14, # 6, 835 – 841 (1998)

[77] Ericksen J. L. *Anisotropic fluids.* Arch. Rat. Mech. & Analysis, v. 4, 231 – 237 (1960).

[78] Ericksen J. L. *Conservation laws for liquid crystals.* Trans. Soc. Rheol. V.5. # 1. 23 – 34 (1961).

[79] Ericksen J. L. *Orientation induced by flow.* Trans. Soc. of Rhelogy, v. 6, 275 – 291 (1962).

[80] Eringen A.C., Chang T.S. *A micropolar description of hydrodynamic turbulence.* In: Recent Advances in Engineering Science. V. 5, part 2. Gordon & Breach, New York, 1 – 8 (1970).

[81] Eringen A.C. *Micromorphic description of turbulent channel flow.* J. Math. Anal. & Appl., v. 39, # 1, 253 – 266 (1972).

[82] Eskinazi S., Yeh H. *An investigation of fully developed turbulent flows in a curved channel.* J. Aeron. Sci., v. 23, # 1, 23 – 34 (1956).

[83] Eskinazi S., Erian F.F. *Energy reversal in turbulent flows.* Phys. Fluids, v. 12, # 10, (1969).

[84] Fedorov K.N. *Selected Works in Physical Oceanology.* Leningrad: Gidrometeoizdat, 162 – 173 (1991).

[85] Felzenbaum A.L. *Dynamics of see streams.* Advances (Itogi) in Science and Technology. Hydromechanics, 97 – 338. Moscow: VINITI (1970).

[86] Fenton D.L., Stukel J.J. *Flow of a particulate suspension in the wake of a circular cylinder.* Int. J. Multiphase Flow, v.3, # 2 (1976).

[87] Ferrari C. *On the differential equations of turbulent flow.* In: Continuum Mechanics and Related Problems of Analysis. (The N.I. Muskhelishvili 80th Anniversary). Moscow, Nauka (1972).

[88] Ferziger J.H. *Large eddy numerical simulation of turbulent flows.* AIAA J., v. 15, # 9, 1261 – 1267 (1977).

[89] *Free Turbulent Shear Flows.* Proc. Conf. at NASA Langley Research Center, Hampton, Virginia, July 20-21, 1972; v. 2 - *Summary of Data.* Washington D.C., NASA SP-321 (1973).

[90] Frenkel Ya. I. *Kinetic Theory of Fluids.* Oxford University Press. (1946); Dover, New York (1955).

[91] Fridman A.A., Keller L.V. *Differentialgleichungen fur die turbulente Bewegung einer kompressibelen Flussigkeit.* Proc. First Intern. Congress Appl. Mech., Delft, 395 – 405. 1925; In: A. A .Fridman. Collected works. Moscow: Nauka (1966).

[92] Frost W., Moulden T. H., Eds. *Handbook of Turbulence,* v. 1, Plenum, New York (1977).

[93] Gence J.N. *Homogeneous turbulence.* Annual Review Fluid Dynamics, v. 16, 201 – 222 (1983).

[94] Gledzer E.B., Dolzhansky F.V., Obukhov A.M. *Systems of Hydrodynamic Types and Their Applications.* Moscow: Nauka, 368 pp. (1981).

[95] Goldshtick M.A. *Vortex Flows.* Novosibirsk: Nauka, 366 pp (1981).

[96] Goldstein S., ed. *Modern Development in Fluid Dynamics.* Fluid Motion Panel. Aeronautical Research Committee. London (1943).

[97] Gorodtzov V. A. *Degeneration of fluid turbulence with internal rotation.* J. Appl. Mech. Tech. Phys., # 3 (1967).

[98] Govindaraju S.P., Saffman P.G. *Flow in a turbulent trailing vortex.* Physics of Fluids. V.14. No 10. PP. 2074-2080 (1971).

[99] Gray D.F., Linsky J.L., Eds. *Stellar Turbulence.* Berlin: Springer (1980).

[100] Gurzhienko H.A. *Action of liquid viscosity onto laws of turbulent stream inside straight tunnel with smooth walls.* Proc. TSAGI, # 303. Moscow (1936).

[101] Gurzhienko H.A. *Account for viscosity within the Karman's theory of turbulence.* Proc. TSAGI, # 322. Moscow (1937).

[102] Gurzhienko H.A. *On steady turbulent stream inside conical diffusers with small angles of opening.* Proc. TSAGI, # 462. Moscow (1939).

[103] Haken H. *Advanced Synergetics.* Berlin: Springer-Verlag (1978).

[104] Happel J., Brenner H. *Hydrodynamics of the Small Reynolds's Numbers.* Leyden: Noordhoff (1973).

[105] Head M. R., Bandyopadhyay P. *New aspects of turbulent structure.* J. Fluid Mech., v. 107, 297 – 337 (1981).

[106] Heinloo J. *Phenomenological Mechanics of turbulent Flows.* Tallin: Valgus (1984).

[107] Heisin D.E. *On derivation of averaged equations of glacial cover dynamics with account of asymmetry of stress tensor.* Trans. USSR Acad. Sci. (Izvestia) Physics of Atmosphere & Ocean, v. 13, # 8 (1977).

[108] Hellerman S. *An adapted estimate of the wind stress on the World Ocean.* Monthly Weather Rev., v. 95, # 9 (1967).

[109] Hills R., Roberts P. *Super-fluid mechanics under high concentration of eddy's lines.* Archive for Rational Mechanics and Analysis, v. 66, # 1, 43 – 71 (1977).

[110] Hinze J. O. *Turbulence. Mechanism and Theory.* New York: McGraw (1959).

[111] Hinze J.O. *Turbulent flow regions with shear stress and mean velocity gradient of opposite sign.* Appl. Sci. Res., v. 22, 163 –175 (1970).

[112] Howard L.N. *Divergence formulas involving vorticity.* Arch. Rational Mech. Anal., v. 1, 113-123 (1958).

[113] Hsieh R. *Conductive micro-magnetic fluid.* The 4th Bulgaria Congress on Mechanics. Varna, 957 – 963 (1981).

[114] Ibbetson A., Tritton D.J. *Experiments on turbulence in a rotating fluid.* J. Fluid Mech., v.68, pt 4, 639 – 672 (1975).

[115] Immich H., *Impulsive motion of a suspension: effect of antisymmetric stresses and particle rotation.* Int. J. Multiphase Flow, v. 6, # 5, 441 – 471 (1980).

[116] Iskenderov D.Sh. *Turbulent wake of suspension of rotating solid particles flow past a body.* Applied Mechanics (Kiev), v. 16, # 2 (1980).

[117] Iskenderov D. Sh. *Equation of turbulent flow with intermittency.* Applied Mechanics (Kiev), v. 16, # 12 (1980).

[118] Iskenderov D. Sh., Nikolaevskii V.N. *Turbulent wake of a body and asymmetric hydrodynamics.* Letters Appl. Engng. Sci., v.5, # 3. (1977).

[119] Iskenderov D. Sh., Nikolaevskii V.N. *Mathematical model tornado-like motions with internal eddies.* Proc. USSR Acad. Sci. (DAN), v.315, # 6. (1990).

[120] Iskenderov D. Sh., Nikolaevskii V.N. *Laminar core of atmosphere turbulent eddies.* Proc. USSR Acad. Sci. (DAN), v.319, # 1. (1991).

[121] Ivchenko V.O. *Fluid dynamics on a rotating plane: averaged equations.* Letters Appl. Engng. Sci., v.5, # 6 (1978).

[122] Ivchenko V.O. *On application of asymmetrical mechanics to geophysical hydrodynamics.* Proc. Arctic & Antarctic Inst., v. 357. Leningrad: Gidrometeoizdat (1979).

[123] Ivchenko V.O., Klepikov A.V. *Quasigeostrophic model of ocean circulation with parametric account of mesoscale motions.* Proc. Arctic & Antarctic. Inst., v.387. Leningrad: Gidrometeoizdat (1982).

[124] Ivchenko V.O., Maslovskiy M.I. *On asymmetrical dynamics of glacial cover.* Proc. Arctic & Antarctic Inst., v.357. Leningrad: Gidrometeoizdat (1979).

[125] Jaunzemis W. *Continuum Mechanics*, New York: McMillan (1967).

[126] Joseph D. D. *Stability of Fluid Motions.* Berlin: Springer (1976).

[127] Kagan B.A., Laihtman D.L., Oganesian L.A., Piaskovskiy R.V. *Numerical experiment on seasonal changeableness of global circulation in barotropic ocean.* Trans. (Izvestia) USSR Acad. Sci., Physics of Atmosphere & Ocean, v. 8, # 10, (1972).

[128] Kaloni P.N., De Silva C.N. *Oriented fluids and the rheology of suspensions.* Phys. Fluids, v. 12, # 5, 994 – 999 (1969).

[129] Karman T. *Some aspects of the theory of turbulent motion.* Proc. Intern. Congress for Applied Mechanics, Cambridge (1934).

[130] Karman T., Howarth L. *On statistical theory of isotropic turbulence.* Proc. Roy. Soc., A164, 192 – 215 (1938).

[131] Kiniaki N., Naomichi H. *Turbulent near wake of a flat plate. Pt 1. Incompressible flow.* Bull. Japan Soc. Mech. Engng., v.17, # 108 (1974).

[132] Kline K.A. *Prediction from polar fluid theory, which are independent of spin boundary condition.* Trans. Soc. Rheol., v.19, p.139 (1975).

[133] Kline K.A. *A spin - vorticity relation for unidirectional plane flows of micropolar fluids.* Int. J. Eng. Sci., v, 15, 131 - 134 (1977).

[134] Kline K.A., Carmi Sh. *On the stability of motions of a dilute suspension of a rigid spherical particles.* Bull. del' Acad. Polonaise des Sci., ser. Sci. - Techn., v. 20, # 9 (1972).

[135] Kline K.A., Sandberg T.K. *A polar fluid estimate of relative force.* Acta Mech., v.26, 201 - 222 (1977).

[136] Kochin N.E. *Vector Calculus and First Course in Tensor Calculus.* Leningrad: USSR Acad. Sci. Press, the 7th edition (1951).

[137] Kochin N.E., Kibel I.A., Rose N.V. *Theoretical Hydrodynamics*, v. 1/2. Leningrad – Moscow: OGIZ (1948).

[138] Kolmogorov A. N. *The local structure of turbulence in incompressible viscous fluid for very large Reynolds's numbers.* Proc. USSR Acad. Sci. (DAN), v. 30, # 4 (1941); "Classic Papers on Statistical Theory", Interscience, New York, 151 – 155 (1961).

[139] Kolmogorov A. N. *On degeneration of isotropic turbulence in an incompressible viscous liquid.* Proc. USSR Acad. Sci. (DAN), v. 31, # 6 (1941); "Classic Papers on Statistical Theory", Interscience, New York, 156 – 158 (1961).

[140] Kolmogorov A. N. *Dissipation of energy in locally isotropic turbulence.* Proc. USSR Acad. Sci. (DAN), v. 32, # 1 (1941); "Classic Papers on Statistical Theory", Interscience, New York, 159 – 161 (1961).

[141] Kolmogorov A.N. *Equations of turbulent motion of incompressible fluid.* Trans. (Izvestia) of USSR Acad. Sci., ser. Physics, v. 6, # 1/2 (1942).

[142] Kolmogorov A.N. *On fragmentation of drops in turbulent flow.* Proc. USSR Acad. Sci. (DAN), v. 66, # 5 (1949).

[143] Kossin J.P., Schubert W.H., Montgomery M.T. *Unstable interactions between a hurricane's primary eyewall and the secondary ring of enhanced vorticity.* J. Atmos. Sci., v. 57, # 24, 3893 – 3917 (2000).

[144] Kovasznay L.S.G. *Structure of the turbulent boundary layer.* Phys. Fluids, v.10, pt 2, 25 - 30 (1967).

[145] Kovasznay L.S.G., Kibens V., Blackwelder R.F. *Large-scale motion in the intermittent region of turbulent boundary layers.* J. Fluid Mech., v. 41, pt 2 (1970).

[146] Krause F., Radler K.-H., *Mean-Field Magneto-Hydrodynamics and Dynamo Theory.* Berlin: Academia-Verlag (1980).

[147] Krasnov Yu. K. *Evolution of tornadoes.* In: Nonlinear Waves: Structures and Bifurcations. Moscow: Nauka, 174 – 189 (1987).

[148] Kruka V., Eskinazi S. *The wall – jet in a moving stream.* J. Fluid Mech., v. 20, pt 4, 555 – 579 (1964).

[149] Kurgansky M.V. *Helicity generation in moist air.* Physics of Atmosphere and Ocean, v. 29, # 4 (1993).

[150] Kurgansky M.V. *Vorticity generation in moist air.* Physics of Atmosphere and Ocean, v. 34, # 2 (1998).

[151] Laihtman D.L., Kagan B.A., Oganesian L.A., Piaskovskiy R.V. *On global circulation in barotropic ocean of variable depth.* Proc. USSR Acad. Sci. (DAN), v.198, # 2 (1971).

[152] Landau L.D. *On the problem of turbulence.* Proc. USSR Acad. Sci. (DAN), v. 44, (1944).

[153] Landau L.D., Lifshits E.M. *Mechanics of Continuum Media,* The 2nd edition. Moscow: Gostekhizdat (1953).

[154] Landau L.D., Lifshits E.M *Theory of Elasticity.* The 4th edition. Moscow: Hauka (1987).

[155] Laufer J. *Investigation of Turbulent Flow in a Two-Dimensional Channel.* NACA TN # 2123 (1950).

[156] Laufer J. *The Structure of Turbulence in Fully Developed Pipe Flow.* NACA Report # 1174 (1955).

[157] Lavrentiev M.A., Shabat B.V. *Problems of Hydrodynamics and their Mathematical Models.* Moscow: Nauka (1977).

[158] Lavrovskiy E.K., Semenova I.P., Slezkin L.N., Fominykh V.V. *Mediterranean lenses – fluid gyroscopes in ocean.* Proc. Russian Acad Sci., (DAN), v. 375. # 1 (2000).

[159] Leonard A. *Energy cascade in large-eddy simulations of turbulent fluid flow.* Advances in Geophysics, v. 18A, 237 – 248 (1974)

[160] Leslie F. M. *Some constitutive equations for liquid crystals.* Arch. Ration. Mech. and Analysis, v. 28. # 4, 265 – 283 (1968).

[161] Levine V.M., Nikolaevskiy V.N. *On volume averaging and continuum theory of elastic media with microstructure.* In: Modern Problems of Mechanics and Aviation. Moscow: Mashinostroenie (1982).

[162] Lewellen D.C., Lewellen W.S., Xia J. *The influence of a local swirl ratio on tornado intensification near the surface.* J. Atmos. Sci., v. 57, 527 – 544 (2000).

[163] Liepmann H.W. *Aspects of the turbulence problem.* J. Appl. Math. Phys. (ZAMP), v. 3, # 5/6 (1952).

[164] Lilly D.K. *On the structure, energetic and propagation of rotating convective storms.* J. Atmos. Sci., v. 43, Part 1: 113 - 125; Part 2: 126 – 140 (1986).

[165] Lin C.C. *On periodically oscillating wake in the Oseen approximation.* In: Studies in Mathematics and Mechanics (Presented to R. Von Mises), 170 – 176, Acad. Press, New York (1954).

[166] Lin C.C., Shu F. H.-S. *Density wave theory of spiral structure.* Astrophysics and General Relativity, v. 2, 236 – 329 (1968).

[167] Listrov A.T. *On model of viscous liquid with unsymmetrical stress tensor.* J. Apple. Math. Mech. (PMM), v.31, # 1 (1967).

[168] Littleton R.A. *The Stability of Rotating Liquid Masses.* Cambridge University Press (1953).

[169] Loitzanskiy L.G. *Mechanics of Fluids and Gases.* Moscow: Nauka (1970).

[170] Lorenz E.N. *Deterministic nonperiodic flow.* J. Atmos. Sci., v.20 (1963).

[171] Lorenz E.N. *The predictability of a flow, which possesses many scales of motion.* Tellus, v. 21, # 3, 289 – 307 (1969).

[172] Lugovzev B.A. *Structure of the turbulent ring vortex in the disappearing viscosity limit.* Proc. USSR Acad. Sci., v. 226, # 3 (1976).

[173] Lumley J.L. *Invariants in turbulent flow.* Phys. Fluids, v. 9, # 11, 2111-2113 (1966).

[174] Lupyan E.A., Mazurov A.A., Rutkevich P.B., Tur A.V. *Large-scale vortex generation under spiral turbulence of convective nature.* JETF, v. 102, # 5 (11), (1992).

[175] Lurie M.V., Dmitriev N.M. *Local model of turbulized medium.* Proc.USSR Acad. Sci. (DAN), v.239, # 1 (1976).

[176] Makarenko V.G., Tarasov V.F. *Experimental model of tornado.* J. Appl. Mech. Techn. Phys., # 5, 115 – 122 (1987).

[177] Marshall J. S., Naghdi P.M. *A thermodynamical theory of turbulence. I. Basic developments.* Phil. Trans. R. Soc. Lond., v. A327, 415 – 448 (1989).

[178] Marshall J. S., Naghdi P.M. *A thermodynamical theory of turbulence. II. Determination of constitutive coefficients and illustrative examples.* Phil. Trans. R. Soc. London, v. A327, 449 – 475 (1989).

[179] Marshall J. S., Naghdi P.M. *Consequences of the second law for a turbulent fluid flow.* Continuum Mech. Thermodynamics, v. 3, 65 - 77 (1991).

[180] Mattioli G.D. *Sur la theorie de la turbulence dans canaux.* C.R. Acad. des Sci., Paris, v.196, # 8 (1933).

[181] Mattioli G.D. *Teoria della turbolenza.* Rend. Acad. Naz. Dei Lincei, v.17, # 13, (1933).

[182] Mattioli G.D. *Teoria Dinamica dei Regimi Fluidi Turbolenti.* Padova: CEDAM, (1937).

[183] Maugin G.A., Eringen A.C. *Deformable magnetically saturated media.* Pt. 1-2, J. Math. Phys., v.13, # 2, 9, (1972).

[184] Maugin G.A., Eringen A.C. *Polarized elastic materials with electronic spin - a relativistic approach.* J. Math. Phys., v.13, # 11 (1972).

[185] Meachem S.P., Flieri G.R., Send U. *Vortices in shear.* Dyn. Atmos. Oceans, v. 14, 333 – 386 (1990).

[186] Migun N.N., Prokhorenko P.P. *Hydrodynamics and Heat Exchange in Gradient Flows of Microstructure Fluid.* Minsk: Nauka i Teknika (1984).

[187] Mindlin R.D. *Micro - structure in linear elasticity.* Archive of Rational Mechanics and Analysis, # 1, 51 – 78 (1964).

[188] Moffat H.K. *The degree of knottedness of tangled vortex lines.* J. Fluid Mech., v. 35, part 1, 117 – 129 (1968).

[189] Moffat H.K. *Turbulent dynamo action at low magnetic Reynolds number.* J. Fluid Mech., v. 41, pt. 2, 435 – 452 (1970).

[190] Moffat H.K. *Magnetic Field Generation in Electrically Conducting Fluids.* Cambridge University Press (1980).

[191] Moin P., Kim J. *The structure of the vorticity field in turbulent channel flow. Part 1. Analysis of instantaneous field and statistical correlators.* J. Fluid Mech., v. 155, 441 – 464 (1985).

[192] Molinari J., Vallaro D. *External influence on hurricane intensity. Part I: outflow layer eddy angular momentum fluxes.* J. Atmos. Sci., v. 46, # 8, 1093 – 1105 (1989).

[193] Monin A.S., Kamenkovich V.N., Kort V.M. *Changeableness of World Ocean.* Leningrad: Gidrometeoizdat (1974).

[194] Monin A.S., Obukhov A.M. *Basic laws of turbulent mixing in a ground layer of atmosphere.* Trans. Geophys. Inst. USSR Acad. Sci., # 24 /151/, 163 – 187 (1954).

[195] Monin A.S., Yaglom A.M. *Statistical hydromechanics. Mechanics of turbulence,* parts 1/2.- Moscow: Fizmatgiz (1965 – 1967).

[196] Montgomery M.T., Vladimirov V.A., Denissenko P.V. *An experimental study on hurricane mesovortices.* J. Fluid Mech., v. 471, 1 - 32 (2002).

[197] Naue H. *Laws of conservation of non-classical hydrodynamics and their application to turbulent channel streams.* Numerical Methods of Continuum Mechanics (Novosibirsk), v. 4, # 1 (1973).

[198] Naue G., Kohlmann J., Schmidt W., Scholz R., Wolf P. *Modellierung und berechnung turbulenter stromungen und anwendungen in der technik.* Teil 1-2, Technische Mech. (1980/1982).

[199] Nee V.W., Kovasznay L.S.G. *Simple phenomenological theory of turbulent shear flows.* Physics of Fluids, v.12, # 3 (1969).

[200] Nemirovskiy Yu. V., Heinloo Ya. L. *On averaged characteristics of movement of turbulized electro-conductive liquid in plane channel in external homogeneous electrical and magnetic fields.* Magn. Hydrodynamics (Riga), # 3 (1976).

[201] Nemirovskiy Yu. V., Heinloo Ya. L. *Local-eddy approximation within description of rotational non-isotropic turbulent flows,* part 1 - 2. Trans. (Izvestia) Sib. Branch USSR Acad. Sci., ser. Techn., # 3 (1978).

[202] Nemirovskiy Y.V., Heinloo J.L. *Local –Rotational Theory of Turbulent Flows.* Novosibirsk University, 92 pp (1980).

[203] Nemirovskiy Y.V., Heinloo J.L. *Rotational-Anisotropic Turbulent Flows in Channels and Tubes.* Novosibirsk University, 76 pp (1982).

[204] Nesterovich N.I. *Averaged equations of turbulent flow of heterogeneous mixtures.* Inst. Theor. Appl. Mech. Siberian Branch USSR Acad. Sci. Preprint # 8-82 of, 37 pp., (1982).

[205] Nemtzov V.B. *Statistical theory of hydrodynamic and kinetic processes in liquid crystals.* Theor. & Math. Physics, v. 25, # 1 (1975).

[206] Nevzgliadov V.N. *Thermodynamics of turbulent systems.* JETF, v.39, # 6/12 (1960).

[207] Nguen Van Diep, Listrov A.T. *On non-isothermal model of non-symmetrical fluids.* Trans. (Izvestia) USSR Acad. Sci., Mech. Fluids & Gases, # 5 (1967).

[208] Nieuwstadt F.T.M., Van Dop H., Eds, *Atmospheric Turbulence and Air Pollution Modeling.* Dordrecht: Riedel (1982).

[209] Nigmatulin R.I. *Foundations of Mechanics of a Heterogeneous Medium.* Moscow: Nauka (1978).

[210] Nigmatulin R.I., Nikolaevskiy V.N. *Diffusion of eddy and conservation of moment of momentum within dynamics of non-polar liquids.* J. Appl. Math. Mech. (PMM), v.34, # 2 (1970).

[211] Nikolaevskiy V.N. *Convective diffusion in porous media.* J. Appl. Math. Mech. (PMM), v. 23, # 6 (1959).

[212] Nikolaevskiy V.N. *Asymmetrical mechanics of continua and averaged description of turbulent flows.* Proc. USSR (DAN), v. 184, # 6 (1969).

[213] Nikolaevskiy V.N. *Asymmetrical mechanics of turbulent flows.* J. Appl. Math. & Mech., (PMM), v. 34, # 3 (1970).

[214] Nikolaevskii V.N. *Asymmetric mechanics and the theory of turbulence.* Archwum Mechaniki Stosowanej, v. 24, # 1, 43 – 51 (1972).

[215] Nikolaevskiy V.N. *Asymmetrical mechanics of turbulent flows. Energy and entropy.* J. Appl. Math. & Mech. (PMM), v. 47, # 1 (1973).

[216] Nikolaevskii V.N. *Asymmetric mechanics of turbulence. The transfer of momentum and vorticity in a wake behind the body.* In: Omaggio a Carlo Ferrari, Libreria Edit. Torino Univ., Levrotto & Bella (1974).

[217] Nikolaevskiy V.N. *Stress tensor and averaging within continuum mechanics.* J. Appl. Math. & Mech. (PMM), v. 39, # 2 (1975).

[218] Nikolaevskii V.N. *On nonlinearity and anisotropy of turbulent viscosity.* Letters Appl. Engng Sci., v.3, 395-404 (1975).

[219] Nikolaevskiy V.N. *Some modern problems of mechanics of multiphase mixture.* In: Modern Problems of Theoretical and Applied Mechanics. Kiev: Naukova Dumka, (1978).

[220] Nikolaevskiy V.N. *On spatial averaging as method of development of mathematical model for media with inherent structure.* Trans. (Izvestia) of Armenian Acad. Sci., Mechanics, v.32, # 4 (1979).

[221] Nikolaevskii V.N. *Short note on a space averaging in continuum mechanics.* Int. J. Multiphase Flow, v.6, # 4 (1980).

[222] Nikolaevskiy V.N. *Spatial averaging and the turbulence theory.* In: Eddies and Waves. Moscow: Mir, 266 – 335 (1984).

[223] Nikolaevskiy V.N. *Condition of generation of fluid element spin at wall stream.* Proc. (Izvestia) of Acad. Sci. of Armenian SSR, Mechanics, v. 41, # 2 (1988).

[224] Nikolaevskii V.N. *Dynamics of viscoelastic media with internal oscillators.* In: Recent Advances in Eng. Sci., Lecture Notes In Engng., # 39, Berlin: Springer-Verlag, 210 – 221 (1989).

[225] Nikolaevskiy V.N., Basniev K.S., Gorbunov A.T., Zotov G. A. *Mechanics of Saturated Porous Media.* Moscow: Nedra, 1970.

[226] Nikolaevskiy V.N., Iskenderov D. Sh., Korzhov E. N. *Turbulent fluid as a continuum with inherent structure.* In: Works of the III All Union Seminar on Models of Mechanics of Continuum Medium. Novosibirsk: Computer. Center Sib. Div. USSR Acad. Sci. (1976).

[227] Novozhilov V.V. *Rheology of steady turbulent flows of incompressible fluid.* Fluid Mechanics, # 3 (1973).

[228] Novozhilov V.V. *Theory of Plane Turbulent Boundary Layer of Incompressible Fluid.* Leningrad: Sudostroenie. 165 pp. (1977).

[229] Ozmidov R.V. *Horizontal Turbulence and Turbulent Exchange in the Ocean.* Moscow: Nauka (1968).

[230] Palmen E., Riehl H. *Budget of angular momentum and energy in tropical cyclones.* J. Meteorology, v. 14, 150 – 292 (1957).

[231] Pauley R.L. *Laboratory measurements of axial pressures in two-celled tornado - like vortices.* J. Atmos. Sci., v. 40, # 20, 3392 – 3399 (1989).

[232] Pekeris C.L., Accad Y. *Solution of Laplace equations for the* M_2 *tide in the World Ocean.* Phil. Roy. Soc. London, v.265, 225-234 (1969).

[233] Perry A. E., Chong M. S. *On the mechanism of wall turbulence.* J. Fluid Mech. V. 119. 173 – 217 (1982).

[234] Perry A. E., Henbest S., Chong M. S. *A theoretical and experimental study of wall turbulence.* J. Fluid Mech. V. 165. 163 – 199 (1986).

[235] Petrosyan L.G. *Some Problems of Fluid Mechanics with Asymmetric Stress Tensor.* Erevan University Press (1984).

[236] Pielke R.A. et al. *Atmospheric vortices.* In: Green S.I., ed. Fluid Vorticies. Chapter 14, 617 – 650. Dordrecht: Kluwer. (1995).

[237] Poincare H. *Theorie des Tourbillions.* Paris: G. Carre, (1893).

[238] Polubarinova – Kochina P. Ya. *On one nonlinear equation for partial derivatives met in the theory of underground flows.* Proc. USSR Acad. Sci. (DAN), v. 63, # 6 (1948).

[239] Prandtl L. *Neuere Ergebnisse der Turbulenzforschung.* V.D. 1, v. 77, # 5 (1933).

[240] Prandtl L. *Fuhrer durch die Stromungslehre.* Braunschweig: Vieweg (1956).

[241] Praturi A.K., Brodkey R.S. *A stereoscopic visual study of coherent structures in turbulent shear flow.* J. Fluid Mech., v.89, pt.2 (1979).

[242] Putterman S.J. *Superfluid Hydrodynamics.* North-Holland, Amsterdam (1974).

[243] Rae W. *Flows with significant orientation effects.* AIAA J. V.14, # 1, 11 – 16 (1976).

[244] Rakhmatulin H.A. *Foundations of gas-dynamics of interpenetrating movements of continuum media.* J. Appl. Math, Mech. (PMM), v. 20, # 2 (1956).

[245] Reynolds O. *On the dynamic theory of incompressible viscous fluids and the determination of the criterion.* Phil. Trans. Roy. Soc, v. 186, 123 – 164 (1893).

[246] Reynolds A.J. *Turbulent Flows in Engineering.* London. Wiley (1974).

[247] Richardson L.F. *The supply of energy from and to atmospheric eddies.* Proc. Roy. Soc. A97, # 687, 354 – 373 (1920).

[248] Robertson H. P. *The invariant theory of isotropic turbulence.* Proc. Camb. Phil. Soc., v, 36, # 2, 209 - 223 (1940).

[249] Robertson J. M. *On turbulent plane Couette flow. Proceedings of the 6th Midwestern Conference on fluid mechanics.* Austin: Univ. of Texas, 169 – 182 (1959).

[250] Rogallo R.S., Moin P. *Numerical simulation of turbulent flows.* Annual Review Fluid Dynamics, v. 16, 99 – 137 (1984).

[251] Rogers M.M., Moin P. *The structure of vorticity field in homogeneous turbulent flows.* J. Fluid Mech., v. 176, 33 – 66 (1987).

[252] Rothfusz L.P., Lilly D.K. *Quantitative and theoretical analyses of an experimental helical vortex.* J. Atmos. Sci., v. 46, # 14, 2265 – 2279 (1989).

[253] Rudiger G. *On negative eddy viscosity in MHD turbulence.* Magnetic Hydrodynamics (Riga), # 1, 3 – 14 (1980).

[254] Rubinow S.I., Keller J.B. *The transverse force on a spinning sphere moving in a viscous fluid.* J. Fluid Mech., v. 11, part 3 (1961).

[255] Saffman P.G. *The lift on a small sphere in a slow shear flow.* J. Fluid Mech., v. 22, pt 2, 385 – 400 (1965).

[256] Saffman P.G. *A model for inhomogeneous turbulent flow.* Proc. Roy. Soc. London, A317, 417 - 433 (1970).

[257] Saffman P.G. *Vortex Dynamics.* Cambridge University Press (1992).

[258] Schlichting H. *Grenzschicht - Theorie.* Karlsruhe: Braun. (1965).

[259] Sedov L.I. *Mechanics of Continuous Media.* Singapore: World Scientific (1991).

[260] Serrin J. *Mathematical Principles of Classical Fluid Mechanics.* Handbuch der Physik (Band VIII/1). Berlin: Springer (1959).

[261] Shahinpoor M. *On the continuum theory of liquid crystals of the nematic type.* Iranian J. Sci. Technol., v. 4, # 4, 111 – 142 (1975).

[262] Shapiro L.J., Montgomery M.T. *A three-dimensional balance theory for rapidly rotating vortices.* J. Atmos. Sci., v. 50, # 19, 3322 – 3335 (1993).

[263] Shroedinger E. *What is Life? The Physical Aspects of the Living Cell.* Dublin Inst. Advanced Studies (1945).

[264] Shliomis M.I. *On hydrodynamics of fluids with internal rotation.* J. Experiment. Theor. Phys. (JETF), v. 51, 258 – 265 (1966).

[265] Shteenbek M., Krause F. *Origin of star and planet magnetic fields as result of turbulent motion of their matter.* Magnetic Hydrodynamics (Riga), # 3 (1967).

[266] Smagorinsky J. *General circulation experiments with the primitive equations.* Monthly Weather Review, v. 91, 99 – 65 (1963).

[267] Sorokin V.S. *On internal friction of fluids and gases with concealed angular impulse.* J. Experiment. Theor. Phys. (JETF), v. 13, 306 – 312 (1943).

[268] Starr V. *Physics of Negative Viscosity Phenomena.* New York: McGraw (1968).

[269] Stojanovic R. *Recent developments in the theory of polar continua.* Wien: Springer (1972).

[270] Straub D., Lauster M. *Angular momentum conservation law and Navier-Stokes theory.* Int. J. Non-Linear Mechanics, v. 29, # 6, 823 – 833 (1994).

[271] Sutton O.G. *Micrometeorology.* New York: McGraw, (1953).

[272] Suyasov V.M. *On nonsymmetrical model of viscous electromagnetic fluid.* J. Appl. Mech. Techn. Physics, # 2 (1970).

[273] Sychov V.V. *Viscous interaction of non-stationary vortex with solid surface.* Fluid Dynamics, # 4 (1989).

[274] Tamm I.E. *Theory of Electricity Foundation.* Moscow - Leningrad: GITTL (1949).

[275] Tang K.K., Welch W.T. *Remarks on Charney's note on geostrophyc turbulence.* J. Atmos. Sci., v.58, 2009 - 2012 (2001).

[276] Tassul J.-L. *Theory of Rotating Stars.* Princeton University Press (1978).

[277] Taylor G.I. *On the dissipation of eddies.* (1918). In: Taylor G.I. Scientific Papers, v. 2, 96 - 101, Cambridge University Press (1960).

[278] Taylor G.I. *The transport of vorticity and heat through fluids in turbulent motion.* (1932). In: Taylor G.I. Scientific Papers, v. 2, 253 - 270, Cambridge University Press (1960).

[279] Taylor G.I. *Statistical theory of turbulence.* Pts I – IV. (1935). In: Taylor G.I. Scientific Papers, v. 2, 288 - 335, Cambridge University Press (1960).

[280] Tilli D.R., Tilli J. *Superfluidity and Superconductivity.* New York: Reinhold (1974).

[281] Townsend A.A. *The fully developed turbulent wake of a circular cylinder.* Austr. J. Sci. Res., A, v.2, # 2 (1949).

[282] Townsend A.A. *Structure of Turbulent Shear Flow.* Cambridge University Press (1956).

[283] Trofimov V.M. *On effect of orientation turbulence properties on heat exchange.* Vestnik of Moscow State Technical University, # 3 (1995).

[284] Trofimov V.M. *Physical effect in the Ranque vortex tubes.* JETP Letters, # 5 (2000).

[285] Trofimov V.M. Shtrekalkin S.I. *On turbulent heat exchange in supersound flows with high local pressure gradient.* Thermophysics of High Temperatures. V. 34, # 2, (1996).

[286] Truesdell C. *The Kinematics of Volrticity.* Bloomington: Indiana University Press, 232 pp (1954).

[287] Truesdell C. *Stages of the development of the concept of stress.- In: Problems of Continuum* Mechanics. (The N.I. Muskhelishvili 70th Aniversary), SIAM, Philadelphia, 556 – 564 (1961).

[288] Truesdell K. *Mechanical bases of diffusion.* J. Chem. Phys., v. 37, # 10, 2336 – 2344 (1962).

[289] Truesdell C. *Six Lectures on Modern Natural Philosophy.* New York: Springer, (1966).

[290] Tsebers A.O. *Internal rotation in hydrodynamics of weakly conducting dielectric suspensions.* Fluid Dynamics, # 2 (1980).

[291] Urpin V. *Kinematic turbulent dynamo in a shear flow.* Geophys. Astrophys. Fluid Dynamics, v. 95, 209 – 284 (2001).

[292] Vainstein S.I., Zeldovich Ya. B., Ruzmaikin A.A. *Turbulent Dynamo in Astrophysics.* Moscow: Nauka (1980).

[293] Volovitskaya Z.I. Mashkova G.B. *On wind profile and turbulence features in a lower 300-meters atmosphere layer.* In: "Studies of lower 300-meters Atmosphere Layer", Moscow: USSR Acad. Sci. Press, 14 – 25 (1963).

[294] Voronov V.A., Ivchenko V.O. *On account for action of mesoscale movements onto large-scale oceanic circulation.* Oceanologia, v.18, # 6 (1978).

[295] Wang C.C. *A new representation theorem for isotropic functions.* Arch. Rational Mech. Anal., v. 36, # 3 (1970).

[296] Wattendorf F.L. *A study of the effect of curvature on fully developed turbulent flow.* Proc. Roy. Soc. London, A148, 565 – 598 (1935).

[297] Xi H.-W., Toral R., Gunton J.D., Tribelsky M.I. *Extensive chaos in the Nikolaevskii model.* Physical Review, v. E62, # 1, R17 – R20 (2000).

[298] Yakimov Y. L. *Tornado and special limit solution of the Navier – Stokes equations.* Fluid Dynamics, # 6 (1988).

[299] Yakimov Y. L. *On class of non-stationary self-similar flows without "essential" singularities and tornado generation mechanism.* Fluid Mechanics, # 4 (1992).

[300] Yanenko N.N., Grigoriev Yu.N., Levinskiy V.B., Shavaliev M. Sh. *Non-equilibrium statistical mechanics of point eddy systems in ideal liquid and its applications to modeling of turbulence.* Novosibnirsk, , Inst. Theor. Appl. Mech., Preprint # 22-82, (1982).

[301] Yantovskiy E.I. *On analogy between transfer from two-dimensional turbulence and orientation magnetization.* Magnetic Hydrodynamics (Riga), # 2 (1974).

[302] Zagustin A.I. *Equations of turbulent fluid flow.* Trans. Voronezh State University, v. X, # 2/3, 7 – 39 (1938).

[303] Zaikovskii V.N., Trofimov V.M. *Mechanics of stratification of turbulent heat transfer in a sound field in the presence of rotational anisotropy of the flow.* JETP Letters, v. 65, # 2 (1997).

[304] Zeldovich Ya. B., Kompaneetz A.S. *On the theory of heat conductivity depending on temperature.* In "Collection dedicated to the 70th Birthday of Academician A.F. Ioffe", Moscow: USSR Acad. Press, 40 – 51 (1950).

[305] Zeldovich Ya. B. , Raizer Yu. P. *Physics of Shock Waves and High Temperature Hydrodynamics Phenomena.* Moscow: PhysMatGiz (1966).

[306] Zhang Z., Eisele K. *On the directional dependence of turbulence properties in anisotropic turbulent flows.* Experiments in Fluids. V. 24. 77 – 82 (1998).

[307] Zhukovskiy N.E. *Eddy theory of frontal resistance* (1919). In: "Works Collection", Moscow: GITTL, v. 4 (1949).

9. INDEX